GARDEN FLOWERS
FROM SEED

GARDEN FLOWERS FROM SEED

AN ILLUSTRATED DICTIONARY

Edited by

Richard Gorer

TREASURE PRESS

First published in Great Britain in 1981 by
Webb & Bower (Publishers) Ltd

Designed by Vic Giolitto

This edition published in 1984 by
Treasure Press
59 Grosvenor Street
London W1

Introduction © 1981 Richard Gorer
Text © 1981 Webb & Bower (Publishers) Ltd

Reprinted 1984

ISBN 0 907812 60 0

Printed in Hong Kong

Half-title page: A colourful window-box display.
Frontispiece: A garden in spring with paths and waterside borders.

CONTENTS

INTRODUCTION
Richard Gorer

According to Greek mythology, civilization was the result of Prometheus' revelation to mankind of the use of fire. In Slavonic legend the Prometheus-figure is Premysl and he influenced civilization by inventing the plough. I know of no legend that names the first discoverer of the fact that plants grow from seed, yet this must have been an even more significant discovery than the plough. After all, there is no point in ploughing the land unless you have crops to sow. Nowadays, we all know that plants come from seeds, but in the earliest days of Man's history it must have needed considerable powers of observation to understand that the small hard objects at the ends of flowering plants could in time give rise to further plants. The spread of Man over most of the Earth must be due, to a large extent, to those unknown primitive naturalists who discovered that they could produce plants from sowing seeds.

This fact is so obvious now that we forget what a momentous observation it must originally have been. We collect seeds from our garden or buy them in packets at the seed merchant's without

Foxgloves and hostas form the foreground to this display in a show tent

any thought that we are benefiting from one of the most important discoveries in our history.

There is little difficulty in getting most seeds to germinate. Gardeners have been growing plants for seed from long before the Christian era and there were seedsmen in Britain in the thirteenth century and probably even earlier. This long tradition has had its results on the seeds themselves. If we sow a packet of seeds we can confidently expect them all to come up together; but this is unnatural. In nature, seeds tend to stagger their germination, so that if there were some natural catastrophe, such as a drought or a flood or a forest fire, some seeds would still be left in the ground to appear after the disaster had receded. The saying 'one year's seeding is seven years' weeding' is not fanciful but is almost literally true. Some seeds can lay dormant in the soil for many years until conditions are right for them. When I was young, corn fields scarlet with poppies were common. Now modern herbicides destroy the poppies and leave the cereals so that poppies are rare; but if an old pasture is ploughed up it may well carry a crop of poppies, even though it may have been ten years or more since they were seen in the neighbourhood.

Over the centuries seedsmen have been selecting plants whose seed would germinate regularly and, if possible, rapidly. The easiest way to do this is to take the seeds from those plants that appear first in your seed bed as they presumably carry a gene for early germination which is transmitted to their own seeds. This method of selection works admirably for annuals and biennials, but is less effective with some perennial subjects and is well-nigh impossible for trees and shrubs because of the time factor involved. Peonies, for example, take at least five years to come into flower from seed, while shrubs usually take seven years and trees take ten. We speak glibly of herbaceous perennials as though they all had unlimited life spans, but this is not the case. The perennials that flower the second year after sowing are not usually very long-lived, while plants that take their time, such as Peonies or the Burning Bush, have potentially a much longer life expectancy. The same applies to some shrubs: Brooms and Rock Roses flower in two or three years from seed, but have a short life compared with, for example, Azaleas or Magnolias which will keep you waiting for longer before you see the first flowers. Some seeds have the capacity for staggered germination so firmly fixed that selection seems to have failed completely. The Gentianella, *Gentiana acaulis*, has been a garden ornament for nearly 400 years, but is still extremely tricky to raise from seed. People who do sow the seed of this species reckon that they may have to keep the seed pan for three years before all the seeds will have germinated; they will get some plants the first year, some the second and some the third. Three and a half centuries of selection

Chitted seeds are germinated in the laboratory and sent to the customer ready for potting up.

have not altered this species' behaviour in the slightest.

A seed needs water, oxygen and the correct temperature in order to germinate. As far as germination is concerned, soil is unnecessary. Most of us have grown mustard and cress on blotting paper or a piece of damp flannel. However, germination is only the first stage in growing a plant and to go further some soil is necessary. For the best results, the soil should be open so that air can get between the soil crumbs; quick draining so that the roots do not drown as a result of water staying in the soil and preventing the oxygen from getting to the roots; and light so that the young tender roots can extend easily and nourish the plantlet. Special seed composts have been formulated, some with loam as a base and some that are loamless. The best loam-based one is the John Innes formulation, which is made up of 2 parts sterilized loam, 1 part peat, and 1 part sharp sand. All the parts are measured by bulk. To every 8 gallons (36 litres) of this mixture are added $1\frac{1}{2}$ ounces (42g) of superphosphate and $\frac{3}{4}$ ounce (21g) of ground chalk. It is possible, though expensive, to buy JIS (as the seed compost is abbreviated) already made up and this may be essential if you are unable to obtain a good silky loam. This is usually made by stacking turves for a year, so that they break down and is becoming increasingly hard to obtain. Garden soil is not the same thing, although, since the seedlings usually do not

9

stay long in the seed compost, it is possible to use this as a last resort. Sterilizing the soil is not easy for the amateur. The easiest way is to put it in a low oven at 250°F (130°C) and leave it there until the soil has reached this temperature throughout. This heating process kills all the fungus spores in the soil which can be lethal to seedlings. In theory, this sterilized soil should be stored for three weeks in a container, such as a clean plastic dustbin, as the sterilization not only destroys all the fungus spores but also the beneficial bacteria which need three weeks to re-establish themselves. There are other ways of protecting your seedlings from fungal attacks if you feel that sterilization is too laborious. There are, for example, various fungicidal dusts with which you can treat the seeds. These are usually sold in puffer packs so that you can puff the dust over the seeds. Alternatively, there is a preparation called Cheshunt Compound which dissolves in water. This solution is used to water the pot immediately after sowing. Although it is easy enough to cope with fungal infections, sterilization also kills off the weed seeds present in the soil, and if you have not sterilized the soil you must be sure that you can distinguish between the seedlings you intend to grow and any unwanted weed seedlings. This is something that can only be learnt by experience, although to start with you can always call in a more knowledgeable friend to help you.

Nowadays, many people employ peat-based composts which are purchased already made up and sterile. Everything has a bad side as well as a good one and if these peat-based composts become really desiccated it is by no means easy to get them moist again, so you must keep a careful watch to prevent this from happening. Of course, if the compost dries out when the seedlings are in it you will probably lose them, so naturally you must be watchful at this time. The main problem seems to be the half a bag of compost left over from last season. The easiest solution to this problem is to tip the dry compost into a bowl of water and leave it there for some time, then remove it by hand, squeezing out all the surplus water before putting it into the pots and leaving it for a couple of days before sowing the seed. A more serious objection to using peat-based composts is that the seedlings tend to get a severe check when they are moved into loam-based composts or into the open ground. The plants will eventually overcome this, but it takes time. Of course, if you are growing short-lived pot plants, such as Primulas, you can pot them on into peat-based potting composts, and they will thrive with no checks. Indeed, plants with very thin roots, such as Begonias, will do much better in peat-based than in loam-based composts. Most of us tend to grow plants that are to be planted out in the garden in pots to start with, and for these the loam-based composts are more satisfactory.

With many seeds there is no need to use pots; they can be sown out of doors from the outset. If you are growing many of these, it is worthwhile constructing a proper seed bed. Your aim in doing this is to achieve an open, well-drained medium in which the young roots can develop quickly. The recognized method of making a seed bed is to excavate the designated area to a depth of 18 inches (46cm) and, if your soil is heavy and drains slowly, to put a layer of hardcore, such as broken bricks, at the bottom. Then replace the soil, removing all large stones but leaving small pebbles and attempt to get a fine, crumblike texture. This is best done by incorporating garden compost, but if that is not available peat will serve as a substitute. You can, of course, control the amount of water a plant will receive when it is grown in a pot, but in the open you may have too much rain. If this causes the soil on your seed bed to pan down it can damage your seeds, so your soil must not be too fine and it may well be worthwhile to incorporate plenty of grit in order to keep the texture open. Making a good seed bed is a laborious task, but once made it will serve for many years and certainly will repay your initial trouble. After the first year it is worth scattering some general fertilizer, such as Growmore, over the seed bed at the rate of not more than 1 ounce (25g) per square yard (or square metre) in late February or early March—in any case, at least a fortnight before you start sowing. You will be using this outside seed bed mainly for biennials, such as Wallflowers or Polyanthus, or herbaceous perennials, such as Lupins or Delphiniums. Hardy annuals are usually sown where they are to flower, although many, such as Larkspurs and Cornflowers, will make much better plants if they are treated as biennials and sown outside in the autumn and put in their flowering positions during the following spring.

Before considering the technique of sowing, let us look at the structure of the seed. This is most easily seen in a very large seed, such as a broad bean. If you soak one for a day you should be able to open it fairly easily. The outer part of the seed is a fairly thin membrane, known as the seed coat to most of us, but as the testa to botanists. Inside the testa is a comparatively large mass of material, which acts as a food reserve, and a tiny embryo plantlet, consisting of a rootlike appendage and a minute bud. With small seeds this embryo is so tiny that magnification is necessary to see it. Flowering plants are divided into two great families, known as dicotyledons and monocotyledons. These can easily be distinguished at the seedling stage: the dicotyledon produces two seed leaves, which are usually different in shape from the adult leaves; the monocotyledon produces a single leaf, which looks very similar to the later adult leaves. Monocotyledons include almost all bulbous plants and normally they take a long time to develop to the flowering stage, although Freesias can be made to flower

Tubs of begonias, petunias and lobelias make a colourful display for a patio or small garden.

within seven months of being sown. In most dicotyledons the food reserves are included in the cotyledons, while in mono-cotyledons it may be in another part of the seed.

When the seed is sown the first thing it does is take in water; no germination is possible before this happens. Some seeds have very hard seed coats which are slow to admit water. Many Sweet Pea growers chip a little of the seed coat away before sowing the seeds and this may speed up germination—although Sweet Peas are not among the worst seeds for slow germination. Many seeds have elaborate chemical mechanisms to ensure that they germinate at the right season. For example, many plants that come from areas like the Mediterranean, where the summers are hot and dry, have a period known as after-ripening. This means that the seeds are not completely ripened when shed and undergo a further ripening after separation from the parent plant. If you grow that attractive little Greek plant *Anemone blanda*, with its bright blue flowers in early spring, you may be lucky enough to find self-sown plants coming up in your garden. If, however, you look for ripe seeds you will never find them. They are shed while they are still green and complete their ripening in the soil. The point of this mechanism is that it prevents the seeds from germinating in the summer, when they would probably dry out before their roots could grow deep enough into the soil. The Anemone, indeed, takes a complete rest during the summer; the leaves die off and the plant remains dormant as a tuber. When the autumn rains come it starts to come back to life, and at the same time the seeds germinate. Many spring bulbs have this mechanism which stops them from germinating until the autumn or winter. Even the Mediterranean area has summer storms and, without this mechanism, many seeds might start growing after a heavy shower, only to be burned up later.

Since plants cannot move, one of their problems is to distribute their seeds as widely as possible. One way they achieve this is by surrounding the seeds with a fleshy, palatable pulp to entice birds to eat them. The birds eat the pulp and the seeds emerge with their droppings, hopefully at some distance from the parent plant. As the seeds pass through the bird, the seed coat is affected by certain chemicals and by the sharp stones in the bird's gizzard, so the seed coat becomes permeable. If you pick hawthorn berries, for example, and remove the seeds and sow them, you may have to wait as long as two years before they will start to grow, but seeds that have passed through a bird's digestive system will tend to germinate the next spring, as they are able to take up water right away.

Unfortunately, some seeds can take in water and swell visibly even after they are dead, so the fact that the seed has taken in water does not automatically mean that it will germinate. What is certain is that if it has not taken up water no plant can be hoped for. Once the seed has taken in water, its next requirement is oxygen. This is present in the atmosphere and there is enough air between soil particles for the seed's purpose. This would not apply if the seed were planted too deep in the earth, so seeds are usually sown shallowly, rarely more than $\frac{1}{2}$ inch (1cm) deep, while very small seeds are either left on the surface of the compost, or barely covered with a layer of sharp sand. With water and oxygen and a reasonable range of temperature (which varies from plant to plant, but 60–70°F (15–21°C) is usually sufficient for most plants), chemical changes start to take place within the seed. The enzymes in the food reserves are dissolved and are transferred to the embryo plant, causing it to expand. First, a rootlike structure, the radicle, emerges. Although this looks like a root, it does not absorb nutrients in the way that true roots do; its purpose is to anchor the young plant firmly in the soil. Once this has been effected, the cotyledons tend to push their way upwards out of the soil, where they unfold in the case of dicotyledons, and elongate in the case of monocotyledons. As they unfurl, many dicots shed the testa which has been protecting the cotyledons in their early stages and these now become green. In the meantime, true roots are being produced from the radicle. Between the cotyledons of dicots, and from the base of monocots, the true leaves are formed and eventually emerge. Sometimes in dicots these are referred to either as the first true leaves, or as the first rough leaves. With fairly sizeable plants the appearance of the first true or rough leaves is usually regarded as a suitable time to remove the seedlings from the seed compost and transfer them to a potting compost.

Some seeds, of which the most notable are Cinerarias and Winter Cherry (*Solanum capsicastrum*), seem to do better if they are sown directly into potting compost, and the same applies to such plants as tomatoes and cucurbits (cucumbers, marrows, etc). Seed composts are adapted to give the seedlings a good start, but contain few nutrients. These are supplied in the potting compost. The John Innes potting compost (JIP) is made up of 7 parts loam, 3 parts peat, 2 parts sharp sand. To every 2-gallon (9-litre) bucketful of this is added 1 ounce (25g) of John Innes Base and $\frac{1}{4}$ ounce (7g) of ground chalk. This serves for plants in 3-inch (8-cm) pots and seed boxes. For plants in 5- or 6-inch (13- or 15-cm) pots the amount of JI Base is doubled and the mixture is known as JIP2, while for pots above 6 inches (15cm) JIP3, with three times the amount of JI Base, is used. The large quantity of loam means that it is not so easy for the young roots to penetrate the soil, but the mixture is much more nutritious and, once established, the young plants soon start to grow rapidly. The peat-based formulations usually have only a single mix regardless of the size of the container, and additional feeds have to be given when larger pot sizes are used.

Pricking out is among the most important operations in raising seedlings. Many people start their seeds in pots and prick out into seed trays. Some of these are only 2–3 inches (5–8cm) deep, which is not really large enough. Trays from 4–6 inches (10–15cm) deep are to be preferred. These should be filled nearly to the rim with potting compost which is lightly firmed down. The seedlings are removed from their pot, avoiding any damage to the roots, and inserted in the tray in rows about 2 inches (5cm) apart in each direction. This is done by making a hole with a stick, inserting the seedling so that it is at the same depth in the soil as it was originally, and then firming the soil around the roots. Seedlings are delicate so do not use undue force at any time during this operation. After the tray is filled, the compost is watered with a watering can with a fine rose attached, so that the seedlings are not knocked over. In spite of all the care you take, the seedlings will have experienced some root damage and they will mark time for a few days while they make that good. Afterwards, they will start to grow and will soon make sizeable plants. If the seedlings have been under cover so far, now is the time to harden them off. The tray is taken outside and if you have a cold frame they are put in this and after a few days the lights are removed. If you have no frame, put them outside in a position with a westerly aspect, or even a northerly one, so that the tender seedlings will not be scorched by fierce sunshine. After three weeks the seedlings will be completely hardened off and can be planted wherever they are to grow in the open ground.

If the seeds are being grown outside all the time the process is simpler. The seeds are sown in drills in the seed bed and, once large enough to handle, they are lined out in a nursery bed, probably at 4 or 6 inches (10 or 15cm) apart. They can then be left until you are ready to put them in their permanent positions. This technique applies particularly to biennials, such as Wallflowers or Polyanthus, which you may be planning to put in beds once the summer bedding is finished. The seeds are usually sown in June and the plants lined out in July, when they can be left until September or October. The exception to this is the Foxglove, which has such minute seeds that they must be started in a pot. With plants from very small seeds, such as Foxgloves, Begonias, or Calceolarias, you will probably have to wait until they have made more than the first pair of rough leaves before they are large enough to handle with safety. If you have very keen eyesight and delicate fingers, you may be able to prick them out earlier; it is really a matter of your ability.

The main point to remember when actually sowing the seeds is that to get the best results you need to sow thinly. Crowded seedlings not only grow more slowly, since they are competing with each other, but are also more liable to succumb to any fungal diseases which may occur. Large seeds can, of course, be inserted individually in pots. Sweet Peas, for example, are usually sown direct into pots, either singly or in groups of three, and any sizeable seed may be spaced out fairly evenly. For smaller seeds the normal procedure is this: fill the pot with seed compost to within $\frac{1}{2}$ inch (1cm) of the rim and firm down. Some people water the compost at this stage and then leave it for twenty-four hours. Tip the amount of seed you think you will need into the palm of your left hand and pick up a pinch with the thumb and forefingers of your right hand. (If you are left-handed, the positions would be reversed.) You will find it quite easy to distribute the seeds thinly over the surface of your pot. With very fine seeds it may be worth mixing the seeds in your palm with a small amount of sharp sand, but take care to mix the seeds and sand well together. Once the seed is sown some people pat the surface of the soil with the palm of their hand to ensure that the seeds are well in contact with the soil. Then cover the seeds with a layer of sharp sand. If silver sand is readily available, this is by far the best, but it is rather expensive. Very fine seeds can be left uncovered and small seeds only need the lightest of coverings. Only the larger seeds need a covering of $\frac{1}{4}$ inch (6mm). If the compost has already been watered, simply moisten the top surface. If you did not water the compost before planting the seeds, do so now; apply the water with a fine rose, but ensure that it penetrates all through the compost.

Some people cover the pot with a sheet of newspaper and a piece of glass; others do nothing. The glass and newspaper prevent evaporation, so that it is unnecessary to water the compost again until the seedlings appear. Once they do, remove the paper and glass so that the seedlings receive all the light available. It is just as effective to envelop the pot in a polythene bag, which has the advantage that it can be left after the seedlings have appeared. However, the bag must be removed and turned inside out about every four days, otherwise too much condensation may occur. If you choose not to cover the seeds in either of these ways watch to see that the compost does not dry out and apply water if it does.

Watering is an essential part of any pot culture. Rain water is always to be preferred, although tap water is an acceptable substitute. It is important that it should be at the same temperature as the atmosphere in which the plants are growing. No harm will occur if it is slightly warmer, but very cold water can check the growth of seedlings by lowering the soil temperature. The simplest way to get the correct temperature is to store a container full of water in your greenhouse, kitchen, or wherever you are raising your seedlings, for at least twenty-four hours before using it. When watering the seedlings, water from above through a very fine rose, but take care not to knock over the seedlings as they rarely recover from this. If, however, some are knocked over, wait a few hours and then gently tease them back into an upright position. You will find a matchstick is an adequate tool for this job. A safer, although longer, method is to hold the pot in a container full of water, with the rim just above the surface of the water. Wait for the water to percolate up through the drainage hole in the pot and, when the surface becomes damp, remove the pot from the water and allow it to drain, before replacing it on the bench. When large enough, the seedlings can be watered with less care, but they must still be watched fairly carefully while this operation takes place. It is important that the compost is moist but not sodden. This is best accomplished by giving a good watering when required and then letting the compost dry out before the next watering.

So far, it has been possible to give general advice, but this is less easy when it comes to the matter of the temperatures required for the best germination. It should be fairly clear that plants from the tropics and sub-tropics require higher temperatures than those from more temperate climes. Mexican plants, such as Zinnias and

Brightly coloured borders make an effective setting for formal paving.

Scarlet Salvia, although they will spend the summer outdoors quite happily, require quite high temperatures, around 70°F (21°C), to germinate rapidly. Most plants from temperate regions will germinate with a day temperature of 60–65°F (15–18°C) and you will find that the majority of the seeds you require will germinate readily in this temperature range. I say the day temperature, since a fall in temperature at night not only does no harm, but seems definitely to improve performance. This happens in nature and seeds are attuned to this natural fluctuation of temperature. Desert plants, such as Cacti and South African succulents, experience very dramatic changes of temperature between day and night, but in most regions of the world the difference is only a few degrees. In fact, many seeds from temperate climes will germinate quite happily at a temperature of 50°F (10°C), although they may take a little longer to do so. This temperature, however, is about the lowest that most seeds require and it is advisable to delay sowing until you are sure that you can maintain this temperature. An occasional rise above these temperatures owing to sun heat does no harm at all.

Most seeds are sown in early spring and in many cases the earlier you sow the better, provided you can maintain the correct temperature. Situations such as kitchen window-sills or any well-lit place in a centrally heated house will do admirably. Of course, a heated greenhouse is best, but they are expensive to run and a perfectly acceptable substitute is a small electric propagator case.

Seedsmen sometimes supply plantlets by post. These are F₁ hybrid geraniums 'Orbit'.

If you can maintain the correct temperatures you can sow your seeds in February or even late January. If you have to wait until April when, hopefully, the correct temperatures will occur naturally, not a great deal is lost. It is far better for the seedlings to make steady growth than for them to start with a rush and then receive a check; the later sown seeds will often catch up the earlier sown ones. In fact, a good motto for the seed grower is not to be in a hurry. You will get better plants if growth starts a little late and then proceeds without check, than with earlier growth that is later checked by such things as low temperatures. With unheated structures low temperatures are always liable to occur during early spring, when cold and warm spells follow each other in unpredictable succession. Few plants show the effects of unseasonable low temperatures as clearly as the lovely blue Morning Glory. This is a sub-tropical plant which needs a temperature of around 64°F (17°C) to germinate. If it is exposed to low temperatures, the leaves become etiolated and an unhealthy cream colour rather than green; they also become distorted. It grows out of this when warmer conditions are re-established, but you will get much healthier plants if you delay sowing the seeds until the beginning, or even middle, of May. Any check to their growth is bad for young plants and should be guarded against as much as possible.

Most seeds seem to do best if sown in the spring, but there are exceptions. The seeds of spring bulbs, such as Crocus, Scilla, Snowdrop, etc, should be sown in the autumn and left outside until their soil has been frosted. They can then be left or brought under cover and they will germinate at the same time as the mature bulbs are producing their leaves. There are some very oily seeds that should be sown as soon as they are ripe. Notable among these are the Christmas Rose and other *Helleborus*, and Magnolia. These are not normally available from seed merchants, but you may get seeds from your own or your friends' gardens. Although they must be put into soil as soon as they are ripe, usually they will not germinate until the following spring. If they are not sown right away they may not germinate at all and in any case will take an unconscionably long time to do so.

Many plants for the rock garden need the action of frosts on the seeds to induce germination and these, too, are best sown in autumn and left outdoors during the winter. This does not seem to apply to plants like Aubrieta, but most of the choicer Alpines require this cold treatment, which is not surprising when you remember that in their native mountains it would not benefit them to germinate before the snows have melted.

It is a good working principle that seeds that ripen in the autumn will come to no harm if they are sown when ripe and the frost is allowed to penetrate the soil in which they are sown. Many such seeds have a built-in germination inhibitor, which is removed by the frost. I refer, of course, to such plants as grow naturally where winter frosts are to be expected. Plants from warmer climes, which have to be treated as half-hardy, such as Zinnias or China Asters, would almost certainly be damaged if exposed to frost. Such plants are native to sub-tropical regions, where frost is unknown.

A number of hardy annuals will make larger plants if they are sown in early autumn, usually in early September, but August in northern Britain. They can then make a plant large enough to survive the winter, which will start to grow vigorously during the following spring. Plants such as Larkspurs and Cornflowers do exceptionally well under this treatment. They are sown in the seed bed and planted out where they are to flower at the end of March.

It has already been suggested that hardy biennials are best sown outside in June, but the seeds of Foxgloves are so minute that they should be started in a pot, pricked out into boxes and then lined out. Poppies also have very minute seeds, but they make a long taproot (a solid fleshy root like a small carrot) and plants with taproots rarely survive transplanting, so they must be sown where they are to flower and the seedlings thinned out as soon as they are large enough to distinguish. If they are not thinned out, so that there is at least 9 inches (23cm) between the plants, they will make very small plants and will come rapidly into flower, producing only one or two rather miserable flowers and will die immediately after blooming.

There are a number of half-hardy biennials which are popular, such as the greenhouse Primulas, Cinerarias, Calceolarias. These are usually sown about the end of June and they can be sown outside at that time of year or started in pots. As soon as they are large enough to handle, they are put into small pots individually and potted on as these pots become full of roots. Primulas usually end in 5-inch (13-cm) pots, Cinerarias and Calceolarias in 6-inch (15-cm) ones. The pots are usually left outside in a shady position until mid-September, when they are brought under cover. A slightly different procedure is necessary for Schizanthus, sometimes known as Poor Man's Orchid. The seeds are not sown until August, and are potted up as soon as the seedlings are large enough to handle. The secret is to keep them growing, so that they are moved into larger pots as soon as a root is seen at the edge of the soil ball. This means that you have to inspect the soil by turning them out of their pots about every two weeks. If you continue to pot them on, you can end up with a huge plant in an 8- or 9-inch (20- or 23-cm) pot. Such a plant will be a mass of flower in April. If you do not pot them on very rapidly they start to come into flower in the late autumn and are disappointing. During the winter they should be kept frost-free, but they do not like much heat during the winter and 50°F (10°C) is the ideal temperature. The other half-hardy biennials will be reasonably happy at this temperature, but actually prefer it slightly warmer.

Cacti and other succulent plants are becoming increasingly popular and for them a special compost is necessary. There are a few variants, given below, but they all require crushed brick, which is not always available. It can be replaced by crushed clinker, but this must be weathered for six months outdoors before it is safe to use. The recommended composts are: equal quantities of coarse loam, grit, and crushed bricks and mortar rubble; 2 parts (by bulk) loam to 1 part each of peat, grit and crushed brick; for the South African succulents a rather richer mixture of 3 parts loam, 3 parts grit, 2 parts peat and ½ part crushed brick seems to give better results.

These composts are put into boxes and topped with sharp sand. The seeds are then sown thinly on the surface and watered in. If the seeds are large they can be lightly covered with sharp sand, but generally they are rather small and are left uncovered. A temperature of 64–70°F (18–21°C) gives the most rapid germination and the seeds should be sown as soon as this temperature range can be maintained. Young Cacti resent root disturbance and are usually left in their boxes for twelve or eighteen months, so they will have become quite sizeable by the time they are potted up. It is not usually safe to do this before at least twelve months have elapsed, so thin sowing is essential. If the seedlings are too crowded they must be thinned out, but since you get very little cactus seed in a packet and it is expensive, it is advisable to sow thinly. Incidentally, although Cacti and succulents can withstand long dry periods, the seed compost should be kept moist throughout the spring and summer and must not be allowed to dry out completely during the autumn and winter, although it should be on the dry side during these seasons.

Bulbs grown from seed need rather special treatment. The seedlings will usually die if transplanted when young, so you can reckon to keep them in their pots for one or two years. This means that they are sown in a potting compost, not a seed compost, and JIP1 or a peat-based compost will do equally well. You need to work with pots, as boxes are too shallow for bulbs and a 5- or 6-inch (13–15-cm) pot is usually the most satisfactory. Most bulbs have large seeds, so they can be spaced out with ease. I have already noted that most spring-flowering bulbs must be sown in the autumn (I refer here only to hardy bulbs), while summer-flowering bulbs are sown in the spring. With most small bulbs there is usually an interval of three years between sowing the seed and seeing your first flowers, but some South American species are much quicker, while Tulips and Daffodils take longer. Tigridias tend to behave like half-hardy perennials and will often flower the year after sowing, while the modern Freesia will flower in seven months. Freesias are sown at the rate of twelve seeds to a 6-inch (15-cm) pot and usually germinate in three weeks. They are kept growing and about mid-June they are brought out of doors into a well-lit situation which does not get much direct sunlight. They are brought back under cover about the middle of September, but they require ample air and do not relish much artificial heat. Indeed, if there is too high a temperature during the winter the plants may go blind; they will be perfectly happy as long as they are not actually exposed to frost. By sowing pots in succession you can have Freesias all through the winter. A well-glazed structure is, however, necessary — dwellings are usually too dark.

Other bulbs are left in their pots for two years at least, although they are dried out when the leaves have gone. There is one exception that should be noted. Peruvian Lilies (*Alstromeria*) are sown in the usual way, but are potted into 3-inch (8-cm) pots as soon as they are large enough to handle, as they make very brittle tubers which are hard to move. It may be possible to shift them into 5-inch (13-cm) pots during the summer. When the leaves die down in the autumn, they may be turned out of their pots and planted really deeply where they are to flower. Some will even flower the second year, but the third year is more likely. This plant seems to appreciate being planted at least 6 inches (15cm) below the soil and does not thrive on heavy soil.

Bulbs like Tigridias and their allies are kept in pots for the first year, but can be treated like Gladioli the next year, being planted out when all risk of frost has gone, and lifted and dried off in the autumn. Hardy bulbs will probably grow for most of their first year, but the foliage will eventually die down and the pots should then be kept on the dry side until the autumn, when it should be kept moist. In the second year the leaves die down somewhat earlier. With smaller bulbs, such as Snowdrops and Scillas, it is probably most convenient to leave them in their pots with the soil on the dry side until early autumn, when they can be planted out where they are to flower. If you think the young bulbs look too small you can repot them and keep them under control for a further year. Larger bulbs, such as Daffodils, Tulips, and Hyacinths, take much longer to reach their flowering stage. Tulips are particularly tiresome as they fritter away their energy in making side bulbs and may keep you waiting for years before they flower. Narcissi seem to do best if the soil is always damp. They have a very short resting period and start making new roots long before any leaves are visible. They are best planted out in the second or third July after germinating.

As so often happens, Lilies prove exceptions to every rule. Some Lilies, including most of those from North America, Europe, and the Caucasus, produce roots only during their first season and you may well think you are having no success. Assume, however, that they are growing successfully and keep the compost moist. Keep the pots outside during the winter, but out of the range of vermin as, if you are lucky, there will be small bulbils in the soil. After the requisite cold period, leaves will appear in the following year. These leaves will be frost tender, so it is advisable to bring the pots indoors from March until the risk of frosts has gone. Most of the Asian Lilies will produce their leaves and roots in the same year and the Regal Lily may even flower the second year after sowing. There are two tender Lilies, *L. formosanum* and *L. philippinense*, which are almost certain to flower in their second year, but they have to be kept under cover during the winter. They are ready to be put into separate pots after their first season. Otherwise, the Asian Lilies are usually ready to plant out after two years in the pot, while the others require an extra year. Incidentally, Peonies have the same irritating habit of only making roots the first year and waiting for the second season before they produce leaves. They are very slow to come into flower, taking from five to seven years from seed.

Half-hardy bulbs, like the South African Ixias, Gladioli, Babianas, Sparaxis, and so on, require the same treatment as

Freesias to start with, but they are dried off in the winter and started again in the spring. If they are too crowded in their pots, they can be repotted while they are dormant. After the second year the majority will be of flowering size and can either be flowered in their pots or planted outside in the spring, and lifted after flowering has ceased and stored like Gladioli for the winter. In very mild districts they will survive outside all the year round. With the current price of bulbs it is economically sound to curb your impatience and grow them from seed—provided you can obtain the seed.

Trees and shrubs may take years to flower, but there are exceptions among the shrubs. Brooms, for example will usually flower in their third season. They have a taproot which must not be damaged and it is best to sow two or three seeds in a 3-inch (8-cm) pot. If they all germinate, reduce them to a single plant in each pot. This plant can be left in its pot until the autumn and then put where it is to flower, or it can be potted on into a larger pot and grown into a larger specimen before planting out. Planting out can be delayed until the second year. The seeds are sown in spring in the normal way but some people find that if they are put in hot water and left for twenty-four hours they germinate more readily. The Rock Roses (*Cistus* spp) also soon make flowering-size shrubs. Most of them are somewhat tender and may be killed in severe winters, but in milder districts they are usually reliable and their showy flowers, although short-lived, are produced over a six-week period in midsummer. *Abutilon vitifolium* grows very rapidly from seed and many plants flower in their second season. Both this and the Rock Roses are best sown in spring in the usual way and should be potted up individually as soon as they are large enough. Later, they can be either potted on or planted out. Most trees and shrubs come fairly readily from seed, but trees which rely on having their fruits eaten by birds and the seeds later voided usually take a long time to germinate. Hawthorn seeds, for example, if sown in the autumn will probably not germinate until eighteen months have elapsed and many Roses are equally slow to appear. However, the dwarf Fairy Roses will usually germinate quickly. Berberis, although they have seeds surrounded by pulp, also usually germinate fairly quickly. They are sown in the autumn, left outside during the winter and brought into warmer conditions in February. Like all pulpy seeds, the pulp should be removed before sowing, a fairly messy and tedious job. You should get good germination in the spring and although Berberis make little growth during the first year, they accelerate in later seasons and will usually flower in the fifth year.

Trees are a different proposition. The exceptions are the few hardy Eucalyptus, which grow extremely fast. The seeds are sown in the spring, preferably in February in a temperature during the day of 60°F (15°C). The seeds are minute and are just pressed into the top of the compost. As soon as they are large enough to handle, they are potted up individually; plastic or paper pots should be used which can accommodate at least 4 inches (10cm) of soil. The soil should have a high nitrogen content, for example, JIP2. The potting on is done when the second pair of leaves appears, as by this time there is much root action underground. After about ten days, when it is clear that the transplanted seedlings are growing again, they are gradually hardened off. They are put outside in their final positions as soon as the soil has warmed up, usually in mid-June. They grow so rapidly that they will require staking, as the growths are long and whippy. For the first two winters they may be frost tender and it is advisable to tie straw or sacking around the base of the young tree as a protection against very severe frost. After that, there is usually no problem. *E. gunnii*, the Cider Gum, which seems to be hardy under any circumstances, is the most frequently grown.

Other trees grow far more slowly. They are best sown outdoors in autumn and many can be sown in the open ground. Once they are large enough to handle, they are lined out in a nursery bed, initially about 12 inches (30cm) apart. It may not be convenient to put them in their permanent positions while they are still young plants, in which case they should be transplanted every autumn (after the first), so as to prevent them from throwing down long anchoring roots, which would make subsequent movement difficult and hazardous. The width between the young trees should also be increased as they grow larger. Once they are sufficiently large they should be put in their permanent positions. There are a few exceptions to the autumn-sowing rule. Birch seeds will germinate only around midsummer and Willow seeds must be sown almost as soon as they are ripe. A large number of shrubs and small trees will flower after five years, so if you are not too impatient there is much to be said for sowing the seeds of these species.

Finally, a word about making a lawn from seed. This is far cheaper than laying turf, but it is by no means a trouble-free operation. The approved times for sowing are early April or early September and, personally, I prefer autumn sowing. The soil must be well prepared and if it is badly drained it should be dug out to a depth of 12 inches (30cm) and a layer of hardcore or large clinker should be put down. The soil is then replaced, lightened with grit if it is too heavy, and enriched with leaf mould or garden compost if it is too light, and left to settle. If this operation has been done in the spring it should be possible to get rid of both annual and perennial weeds during the summer. Before sowing, the ground should be worked to a fine tilth. The seed should be

Beds of sweet peas: 'Swan Lake' (white), 'Elisabeth Collins' (pink) and 'Excelsior' (crimson).

applied evenly and not too thickly on a calm day. The usual recommendation is 1½oz (42g) of seed to the square yard (square metre). The ground is then lightly raked over and rolled. Germination is brisk, but should there be an unseasonable dry spell, the young grasses should be watered with a sprinkler.

If you intend to sow grass seed during the spring, you must do all the preparation in the autumn. This has the advantage that the soil will have reached its natural level during the winter, but you may not be able to get rid of any perennial weeds that are present in the soil. Whenever you sow, it is advisable to scatter a good lawn fertilizer at the rate of about 2oz (56g) per square yard (square metre) a fortnight before you plan to sow the grass seed.

Once your lawn is fairly well established, rolling is said to improve its performance and this should be done prior to the first mowing. All reputable seedsmen provide good lawn mixtures of various grass seeds, suitable for various conditions. Birds can be a nuisance while the grass seeds are growing and they can be discouraged by putting the seeds in a paper bag with some red lead and shaking them up together before planting, or a zareba of black cotton strung on small sticks can be laid over the sown portion.

No gardening is trouble-free, but you will get good results with a minimum of work if you grow most of your plants from seed. Apart from getting free plants and cuttings from your neighbours, it is the cheapest method of stocking the garden. There is nothing better.

The information given in this book applies to the United Kingdom and countries in the northern hemisphere. However, all the information can be applied to the southern hemisphere provided that the sowing times are advanced by six months. Thus, if in the text you are advised to sow seeds in April, you must sow them in October in the southern hemisphere. Many parts of South Africa and Australia have much higher temperatures than those experienced in the UK and in these areas plants described as half-hardy or even as cool greenhouse specimens may prove to be hardy out of doors. On the other hand, parts of the tropics and sub-tropics may be too hot for some of the seeds mentioned to germinate satisfactorily. Delphiniums, for example, rarely germinate in hot conditions, unless, like the Californian and Persian species, they are native to warm parts of the world.

Note: Unless otherwise stated, the seeds should always be sown in seed compost.

ALPHABETICAL LIST OF POPULAR SPECIES

Acacia dealbata
MIMOSA, SILVER WATTLE

Half-hardy tree
Native to Australia

This plant has fernlike, grey-green foliage, which is covered with silvery down when young, and fragrant fluffy, ball-shaped yellow flowers. In very mild areas it may be grown out of doors against a south-facing wall, but it will not tolerate much frost and does better in a cool greenhouse. Acacia may be treated also as a shrub and grown in a large pot, in which case it reaches a height of about 36–48 inches (91–122cm).

Before sowing the seeds, put them in a bowl and pour boiling water over them. Leave them to soak for 12 hours. Sow each seed in a 3-inch (8-cm) pot during the spring at a temperature of 60–68°F (15–20°C). Germination is sporadic, taking up to 35 days. Pot the young plants on as necessary.

Acacia requires plenty of light and as much ventilation as possible throughout the year. Water the plants moderately during the autumn and winter and freely during the spring and summer. Repot established plants every other year during March. Feed the plants with liquid manure at fortnightly intervals May–August.

When grown from seed, Acacia will flower 3–4 years after sowing. The flowers bloom late winter–early spring.

Achillea
YARROW

Hardy perennial

A. filipendula
Native to the Caucasus

This species has mid-green leaves and 4–6 inch (10–15cm) wide clusters of yellow flowers. It grows to a height of 36–48 inches (91–122cm). It requires a sunny position and well-drained soil, and may be used as a border plant. It is popular with flower arrangers as the flowers may be used fresh during the flowering season and dried for winter decorations.

Sow the seeds April–June at a depth of $\frac{1}{4}$ inch (6mm) in a seed bed. Germination takes 14–21 days. When the seedlings are large enough to handle, thin them out to 12 inches (30cm) apart and grow them on until the autumn. Transfer the plants to their permanent positions October–March, spacing them far enough apart to allow for their spread of up to 48 inches (122cm).

The flowers bloom July–September.

A. millefolium
Native to Great Britain

This species has dark-green leaves and 4-inch (10-cm) wide clusters of white to rose-pink flowers. It grows to a

Alchillea millefolium roseum.

height of 30 inches (76cm). It requires the same conditions as *A. filipendula*.

The sowing instructions are the same as those for *A. filipendula*, but when the plants are transferred to their permanent positions they need not be spaced as far apart because they spread only to about 15 inches (38cm).

The flowers bloom June–September.

A. ptarmica
SNEEZEWORT
Native to Great Britain

This species has mid-green leaves and white daisylike flowers. It grows to a height of 30 inches (76cm). It requires the same conditions as *A. filipendula*.

The sowing instructions are the same as those for *A. filipendula*, but like *A. millefolium*, its spread is only about 15 inches (38cm).

The flowers bloom July–September.

Acroclinium roseum see
Helipterum roseum

Adonis aestivalis
PHEASANT'S EYE

Hardy annual
Native to Europe

This plant has deep-crimson cup-shaped flowers with feathery foliage. It grows to a height of 12 inches (30cm). It thrives in either a sunny or partially shaded position and prefers a light well-drained soil.

Sow the seeds March–May or during September in the flowering position at a depth of $\frac{1}{4}$ inch (6mm). Germination takes 14–21 days. When the seedlings are large enough to handle, thin them out to 12 inches (30cm) apart.

A September sowing will produce flowers early the following summer. The flowers bloom from June onwards.

Ageratum
Half-hardy annual
Native to Mexico

This dwarf plant has clusters of bluish-mauve flowers. It grows to a height of 5–12 inches (13–30cm), depending on the variety. It requires a sunny sheltered site and a moist soil. Ageratum is usually grown for use in summer bedding schemes.

Sow the seeds February–March at a temperature of 50–60°F (10–15°C) and at a depth of $\frac{1}{8}$ inch (3mm). Germination takes 10–14 days. When the seedlings are large enough to handle, prick them out into boxes and harden them off when all risk of frost has passed. Transfer the plants to their

Adonis aestivalis.

Ageratum 'Blue Surf'.

Ageratum 'Blue Blazer'.

Ageratum 'Spindrift'.
Ageratum 'Blue Danube'.

flowering positions during May, spacing them 6–12 inches (15–30cm) apart, depending on the variety. *NB* As Ageratum seeds are minute take care to sow them thinly.

The flowers bloom June–September.

Agrimonia eupatoria
AGRIMONY, COCKLEBUR

Hardy perennial
Native to northern Europe

The leaves are finely divided and fernlike; the yellow flowers bloom early in the summer. The hooked seed pods gave rise to the plant's name of Cocklebur. It grows to a height of 24–36 inches (61–91cm). Agrimony grows on any ordinary garden soil and tolerates partial shade. However, it does better in a sunny position.

Sow the seeds in the flowering site during April at a depth of $\frac{1}{4}$ inch (6mm). When the seedlings are large enough to handle, thin them out to 7–9 inches (18–23cm) apart.

Agrimony is a dye plant used to produce a yellow colour.

Agrostemma coronaria see *Lychnis coronaria*

Allium schoenoprasum
CHIVES

Hardy perennial
Native to Europe and northern Asia

The leaves are spearlike and taste mildly of onion. The rose-pink flowers are arranged in dense round heads. The plant grows to a height of 6–10 inches (15–25cm). It requires an open sunny position and well-drained soil. It may be grown in containers on the kitchen window-sill for use in the winter.

Sow the seeds May–June in open ground at a depth of $\frac{1}{2}$ inch (1cm). Germination takes 14–21 days. When the seedlings are large enough to handle, thin them out to 10 inches (25cm) apart.

Chives may be used to give a mild onion-flavour to soups and salads. They are also used in *sauce tartare* and as *fine herbes*.

Alonsoa warscewiczii
MASK FLOWER

Half-hardy perennial, usually treated as a half-hardy annual
Native to Peru

The dark-green foliage is thickly studded with bright-scarlet flowers which are saucer-shaped. The plant grows to a height of about 9 inches (23cm). It requires a rich, well-drained, loamy soil and a sunny position.

Sow the seeds February–March at a temperature of 50–60°F (10–15°C) and at a depth of $\frac{1}{8}$ inch (3mm). Germination takes 14–21 days. When the seedlings are large enough to

A double-flowered variety of *Althaea rosea*.

A single-flowered variety of *Althaea rosea*.

Althaea rosea 'Majorette': a semi-double variety.

handle, prick them out into boxes and harden them off. Plant them out in their flowering positions during May, spacing them 15 inches (38cm) apart.

The flowers bloom June–October.

Alstroemeria ligtu hybrids
PERUVIAN LILY

Herbaceous perennial
Native to Peru

The flowers are trumpet-shaped and may grow to $1\frac{1}{2}$–2 inches (3–5cm) in width. Colours available include pink, orange, rose, red, and yellow. The plant grows to a height of 30 inches (76cm). It requires a fertile well-drained soil and a sheltered position.

Sow the seeds in a cool greenhouse during March. The seeds should be shown at a depth of $\frac{3}{4}$ inch (18mm) in seed compost. Germination is sporadic, taking up to 60 days. When the seedlings are large enough to handle, prick them out into 3-inch (8-cm) pots. Transfer the plants to their flowering positions during the autumn. Space them 12 inches (30cm) apart and plant them as deeply as possible to protect them from frost.

The flowers bloom June–August.

Althaea rosea
HOLLYHOCK

Hardy perennial usually grown as a hardy biennial
Native to India

This is a traditional cottage-garden flower. The leaves are light green and hairy, and the flower spikes are pink, yellow, red, or white. The flowers may grow to 4 inches (10cm) in diameter. The plant grows to a height of 6 feet (1.8 metres). Double and dwarf varieties are available from seedsmen. Ideally, this plant requires a rich soil and sheltered position. It will require staking.

For flowers the following year, sow the seeds January–February at a temperature of 50–60°F (10–15°C) and at a depth of $\frac{1}{4}$ inch (6mm). Alternatively, sow the seeds May–June in open ground at a depth of $\frac{1}{2}$ inch (1cm). Germination takes 14–21 days. Plant out seedlings grown indoors in April 24 inches (61cm) apart. Thin out seeds sown in open ground to 24 inches (60cm) apart when the plants are large enough to handle.

The flowers bloom July–September.

Alyssum

A. maritimum (*Lobularia maritima*)
Hardy annual
Native to Europe

This species has greyish-green leaves and flowers which are white or various shades of purple. Cultivated varieties are available with pink or rose-coloured flowers. It grows to a height of 3–6 inches (8–15cm). It requires a sunny position and well-drained soil. Its dwarf habit and spreading nature make it especially suitable for use in rockeries or as ground cover.

Alyssum 'Rosie O'Day'.

Alyssum 'Snow Carpet'.

seeds April–May at a depth of $\frac{1}{4}$ inch (6mm) in the flowering site. When the seedlings are large enough to handle, thin them out to allow enough space for them to spread.

The flowers bloom June– September.

A. saxatile

Hardy perennial
Native to eastern Europe

This evergreen species has grey-green foliage and bright golden-yellow flowers. It grows to a height of 9–12 inches (23–30cm). It requires a sunny position and well-drained soil and is suitable for rockeries or borders.

Sow the seeds during March in a cold frame or cool greenhouse at a

Alyssum 'Wonderland'.

Alyssum 'Oriental Night'.

Sow the seeds February–March at a temperature of 50–60°F (10–15°C) and at a depth of $\frac{1}{4}$ inch (6mm). Germination takes 7–10 days. When the seedlings are large enough to handle, prick them out into boxes and harden them off. Transfer the plants to their flowering positions during April, spacing them far enough apart to allow for their spread of 8–12 inches (20–30cm). Alternatively, sow the

depth of $\frac{1}{4}$ inch (6mm) in trays of seed compost. Germination takes 7–10 days. When the seedlings are large enough to handle, prick them out singly into 3-inch (8-cm) pots of potting compost and grow them on until the autumn. Transfer the plants to their permanent positions during September, spacing them 16–24 inches (41–61cm) apart.

The flowers bloom April–June.

Amaranthus caudatus.

Amaranthus

A. caudatus
LOVE-LIES-BLEEDING

Hardy annual
Native to the Tropics

This species has light-green leaves and long drooping tassels of crimson flowers. The variety 'Viridis' has greenish-white flowers. The plant grows to a height of 36–48 inches (91–122cm). It requires a sunny position and a rich, well-cultivated soil. This plant is useful for flower arrangers as the cut flowers last well.

Sow the seeds from the end of March until May in the flowering site at a depth of $\frac{1}{8}$ inch (3mm). Germination takes 14–21 days. When the seedlings are large enough to handle, thin them out to 18 inches (46cm) apart.

The flowers bloom July–October.

A. hypochondriacus
PRINCE'S FEATHER, PYGMY TORCH

Half-hardy annual
Native to tropical America

The foliage is crimson-green and the erect flower spikes are deep crimson.

The plant grows to a height of 48–60 inches (122–152cm). It requires a well-cultivated soil and a sunny position.

Sow the seeds during March at a temperature of 60°F (15°C). Germination takes 14–21 days. When the seedlings are large enough to handle prick them out into boxes and harden them off. Transfer the plants to

Amaranthus hypochondriacus.

their flowering positions during May, spacing them 36 inches (91cm) apart. Alternatively, sow the seeds in the flowering site during April. When the seedlings are large enough to handle, thin them out to 36 inches (91cm) apart.

The flowers bloom August–September.

A. tricolor
JOSEPH'S COAT

Half-hardy annual
Native to India

This species is grown for the sake of its attractive variegated leaves which are coppery-crimson overlaid with bronze, yellow, and green. The plant grows to a height of 24–36 inches

Amaranthus tricolor.

Sow the seeds at any time at a
temperature of 68–78°F (20–25°C).
The seeds should be sown singly in 3-
inch (8-cm) pots of potting compost at
a depth of $\frac{1}{2}$ inch (1cm). Germination
is sporadic, taking 28–42 days.

Plants grown from seed will flower 3
years after sowing. The flowers bloom
September–October.

Ampelopsis veitchii see
Parthenocissus tricuspidata 'Veitchii'

Anchusa

A. azurea *(A. italica)*

Hardy perennial
Native to southern Europe

This tall herbaceous plant has mid-
green leaves and clusters of brilliant-
blue flowers. It grows to a height of
36–60 inches (91–152cm). It succeeds
on any fertile garden soil provided that
it is planted in a sunny position.

Sow the seeds April–June in a seed
bed at a depth of $\frac{1}{2}$ inch (1cm).
Germination takes 14–28 days. When

Anchusa italica 'Suttons Royal Blue'.

(61–91cm). It is usually grown as a pot
plant, but is very striking when used in
outdoor bedding schemes. It requires
a sunny position and a rich, well-
cultivated soil.

Sow the seeds during April at a
temperature of 60–68°F (15–20°C) and
at a depth of $\frac{1}{8}$ inch (3mm).
Germination takes 14–21 days. When
the seedlings are large enough to
handle, prick them out into boxes and
grow them on. Harden off at the end of
May those seedlings that are to be
planted outside. Transfer them to
their flowering position about 10 days
later, spacing them 12–18 inches
(30–46cm) apart. Transfer those
seedlings that are to be grown as pot
plants into 5-inch (13-cm) pots of
potting compost once they are well
developed.

Amaryllis belladonna

Hardy bulb
Native to South Africa

This plant has mid-green foliage and
large trumpet-shaped flowers which
resemble Lilies. It grows to a height of
18–24 inches (46–61cm).

the seedlings are large enough to
handle, thin them out 12 inches
(30cm) apart and grow them on until
the autumn. Transfer the plants to
their permanent positions during
October, spacing them 12–18 inches
(30–46cm) apart.

The flowers bloom June–August.

A. capensis

Hardy annual
Native to South Africa

This species has mid-green leaves and
blue saucer-shaped flowers. It grows
to a height of 18 inches (46cm). It
requires a sunny position and a well-
drained soil and is suitable for use in a
mixed border.

Sow the seeds April–May at a depth
of $\frac{1}{2}$ inch (1cm) in the flowering site.
Germination takes 14–28 days. When
the seedlings are large enough to
handle, thin them out to 6 inches
(15cm) apart.

The flowers bloom July–August.

Anchusa capensis 'Blue Angel'.

Anemone coronaria

POPPY ANEMONE

Hardy perennial
Native to the eastern Mediterranean

This Anemone is available in brilliant
colours which include scarlet, blue,
magenta, and white. It grows to a
height of 6–12 inches (15–30cm). It
requires a moist but well-drained soil
and a partially shaded position.

The seeds are fluffy and need to be
carefully separated before being sown.
Sow the seeds at any time between
January and June at a temperature of
50–60°F (10–15°C) and at a depth of $\frac{1}{8}$
inch (3mm). Germination takes 28–42
days. When the seedlings are large
enough to handle, prick them out into
boxes and harden them off for about
10 days. Then plant them out in their
flowering positions, spacing them

about 4 inches (10cm) apart. Flowers
appear the year after sowing.

The flowers bloom April–July.

Anethum graveolens

DILL

Hardy annual
Native to Europe

The foliage of this herb is feathery and
blue-green in colour. Very small star-

Anemone coronaria.

shaped yellow flowers appear from June to August. It grows to a height of 36 inches (91cm) and thrives on any well-drained soil provided it is in a sunny position.

Sow the seeds April–May in open ground at a depth of $\frac{1}{2}$ inch (1cm). Germination takes 10–14 days. When the seedlings are large enough to handle, thin them out to 3–4 inches (8–10cm) apart.

Dill leaves may be used fresh to garnish salads and may be added to fish dishes and boiled with new potatoes and peas. The seeds are often added to vinegar for pickling gherkins.

Angelica archangelica
ANGELICA

Hardy biennial
Native to Europe

This plant, which has many-flowered decorative heads, may grow to a height of over 6 feet (1.8 metres). It requires a rich soil and a partially shaded position.

Sow the seeds March–April in a seed bed. The seeds should be planted in shallow drills about 12 inches (30cm) apart. Germination takes 14–21 days. When the seedlings are large enough to handle, thin them out to 12 inches (30cm) apart. Leave them in the seed bed until the following March, when they should be transplanted to their permanent positions.

Angelica stems may be candied and the dried leaves may be used in pot-pourri.

Aniseed see Pimpinella anisum

Anthemis nobilis
COMMON CHAMOMILE

Hardy perennial
Native to Europe

The stems lie flat on the ground so that it forms a feathery grey-green mat.

The white daisylike flowers appear July–August. Chamomile grows to a height of 3–10 inches (8–25cm). A light well-drained soil is essential.

Sow the seeds February–March at a temperature of 50–60°F (10–15°C) and at a depth of $\frac{1}{8}$ inch (3mm). Germination takes 14–28 days. Transplant the seedlings to individual pots when the first true leaves appear. Place the plants outside in their permanent positions during May, spacing them 4 inches (10cm) apart if they are to be used as ground cover, or 8–10 inches (20–25cm) apart if they are being grown in rows.

The flowerheads may be used to make a herbal tea.

Anthriscus cerefolium
CHERVIL

Hardy biennial
Native to south-eastern Europe and western Asia

Chervil has bright green, fernlike foliage. White flowers appear June–August in the plant's second year. It grows to a height of 12–18 inches (30–46cm). The plant grows on all types of soil and in sun or partial shade. It also thrives in pots or in window-boxes.

Sow the seeds March–August in shallow drills in open ground. Germination takes 10–21 days. When the seedlings are large enough to handle, thin them out to 6 inches (15cm) apart. When they are large enough to be planted in their permanent positions, space them 12 inches (30cm) apart.

Seeds may be sown in September for use during the winter. Sow these at a temperature of 45–50°F (7–10°C) and ensure that they are well spaced in the seed box.

The leaves may be used in egg and fish dishes, in salads, and in summer soups.

Antirrhinum majus

SNAPDRAGON

Half-hardy annual, hardy annual, or
hardy perennial
Native to the Mediterranean region

Snapdragons are very popular, making
a brilliant display of colour in the
garden. A number of different forms
of this species exist including F_1
hybrids and dwarf varieties. The
foliage is mid-green and the fragrant
flowers, borne on spikes, are tube-
shaped with round lips. The plant
grows to a height of 12–48 inches
(30–122cm), depending on the variety.
It requires a well-cultivated light soil
and a sunny position.

Sow the seeds January–March at a
temperature of 60–68°F (15–20°C) and
at a depth of $\frac{1}{8}$ inch (3mm).
Germination takes 10–14 days. When
the seedlings are large enough to
handle, prick them out into boxes and
harden them off. Transfer the plants to
their flowering positions from late
April onwards, spacing them 9–18
inches (23–46cm) apart, depending on

Antirrhinum majus 'Madame Butterfly'.

Antirrhinum majus 'Triumph Mixed'.

Antirrhinum majus 'Trumpet Serenade'.

the variety. Alternatively, sow the seeds July–September in a seed bed at a depth of $\frac{1}{4}$ inch (6mm). When the seedlings are large enough to handle, thin them out to about 9 inches (23cm) apart and grow them on until the following spring. Transfer them to their flowering positions from late April onwards.

The flowers bloom from July until the first autumn frosts.

Apple of Peru see *Nicandra physaloides*

A long-spurred variety of *Aquilegia*.

Aquilegia
COLUMBINE

Hardy perennial
Native to Europe

This plant has funnel-shaped flowers with a spur on each petal. The most popular form of this genus is the long-spurred hybrid variety. The colours available include blue, pink, white, yellow, and crimson. It grows to a height of 18–24 inches (46–61cm), depending on the variety. Columbine requires a well-drained but moist soil and a sunny or partially-shaded position.

Sow the seeds January–March at a temperature of 50–60°F (10–15°C) and at a depth of $\frac{1}{8}$ inch (3mm). Germination takes 21–35 days. When the seedlings are large enough to handle, prick them out into boxes and harden them off. Transfer the plants to their permanent positions during June, spacing them 12 inches (30cm) apart. Alternatively, sow the seeds April–June in a seed bed at a depth of $\frac{1}{4}$ inch (6mm). When the seedlings are large enough to handle, thin them out to 6 inches (15cm) apart. Transfer the plants to their permanent positions September–October, spacing them 12 inches (30cm) apart.

The flowers bloom June–July.

A. alpina

Hardy perennial
Native to the Alps

This dwarf Aquilegia is brilliant blue and grows to a height of 12 inches (30cm). It requires a well-drained but moist soil and a sunny or partially-shaded position. It is suitable for use in rockeries.

Sow the seeds outdoors April–May at a depth of $\frac{1}{4}$ inch (6mm) in pots of seed compost. Germination takes 21–35 days. When the seedlings are large enough to handle, transfer them to their permanent positions, spacing them far enough apart to allow for their spread of about 12 inches (30cm). Better results may be obtained if the seeds of this species are sown during the autumn, so that they get frozen during the winter. They then germinate during the spring.

The flowers bloom during May.

Arabis alpina
ROCK CRESS, WALL CRESS

Hardy perennial
Native to Europe

This alpine trailing plant has a mass of pink or white cross-shaped flowers. It

Arabis alpina.

of 12–24 inches (30–61cm), depending on the variety. Most soils are suitable, but a sunny position is essential.

Sow the seeds February–March at a temperature of 50–60°F (10–15°C) and at a depth of $\frac{1}{4}$ inch (6mm). Germination takes 14–21 days. When the seedlings are large enough to handle, prick them out into boxes and harden them off. Transfer the plants to their flowering positions during late May, spacing them 12 inches (30cm) apart.

The flowers bloom from June until the first autumn frosts.

grows to a height of 6 inches (15cm). An excellent plant for use in rockeries, it requires a well-drained soil and a sunny position. It also grows on dry walls.

Sow the seeds March–April in a cool greenhouse at a depth of $\frac{1}{8}$ inch (3mm). Germination takes 14–21 days. When the seedlings are large enough to handle, prick them out into boxes. Transfer the plants to their permanent positions during May, spacing them 12 inches (30cm) apart. Alternatively, sow the seeds May–June in their flowering positions at a depth of $\frac{1}{4}$ inch (6mm). When the seedlings are large enough to handle, thin them out to 12 inches (30cm) apart.

The flowers bloom April–June.

Aralia japonica see *Fatsia japonica*

Aralia sieboldii see *Fatsia japonica*

Arctotis × *hybrida*
AFRICAN DAISY

Half-hardy annual
Native to South Africa

This plant has narrow grey-green leaves and daisylike flowers in a wide range of colours. It grows to a height

Arctotis.

Arenaria montana
SANDWORT

Hardy perennial
Native to Europe

This plant is covered with silvery-white flowers during the spring. It grows to a height of 6 inches (15cm). A gritty well-drained soil is ideal. Sandwort is suitable for growing in rockeries and between paving stones.

Sow the seeds May–July at a depth of $\frac{1}{8}$ inch (3mm) in seed compost. Germination takes 14–28 days. When the seedlings are large enough to

Foxgloves (*Digitalis purpurea*) and ornamental grasses add height to this mixed border which also includes roses and tobacco plants (*Nicotiana alata*).

handle, prick them out into boxes and grow them on. Transfer the plants to their permanent positions during September, spacing them 18 inches (46cm) apart.

The flowers bloom May–June.

Armeria maritima
SEA PINK, THRIFT

Hardy perennial
Native to Europe

This plant forms hummocks of grass-like leaves and has flowers in shades of pink, lilac, and cerise. It grows to a height of 6–12 inches (15–30cm). It requires a well-drained soil and a position in full sun. It is suitable for use in rockeries and for edging the front of borders.

Sow the seeds March–April in a greenhouse. Plant them $\frac{1}{8}$ inch (3mm) deep in seed compost. Germination takes 14–28 days. When the seedlings are large enough to handle, prick them out into boxes and grow them on. Transfer the plants to their permanent positions during September, spacing them 12 inches (30cm) apart. Alternatively, plant the seeds during July in the flowering site at a depth of $\frac{1}{4}$ inch (6mm). When the seedlings are large enough to handle, thin them out to 12 inches (30cm) apart.

The flowers bloom May–July.

Asclepias curassavica
BLOODFLOWER, MILKWEED

Half-hardy perennial
Native to the USA

This plant has slender mid-green leaves and large erect clusters of purplish-red and orange-scarlet flowers. It grows to a height of 24 inches (61cm). It requires a sunny but moist site in a slightly acid soil.

Sow the seeds January–March at a temperature of 68–78°F (20–25°C) and at a depth of $\frac{1}{4}$ inch (6mm). Germination takes 14–21 days. When the seedlings are large enough to handle, prick them out into boxes and harden them off. Transfer them to their flowering positions during May, spacing them 12 inches (30cm) apart.

The flowers bloom June–October.

Asparagus, Ornamental

Greenhouse perennial
Native to South Africa

A. medeoloides
SMILAX

This climbing plant grows to a height of 10 feet (3 metres) and must be trained up strings. The dainty foliage is bright green. It should be planted in the greenhouse border.

Sow the seeds at any time at a temperature of 68–78°F (20–25°C). They should be sown in seed compost at a depth of $\frac{1}{2}$ inch (1cm). Germination takes 21–42 days. When the seedlings are large enough to handle, prick them out into boxes of potting compost and then pot them up or transplant them to the greenhouse border.

A. plumosus
ASPARAGUS FERN

This dwarf form of Asparagus is ideal for growing in pots as a houseplant. It has delicate fernlike foliage and grows to a height of 24 inches (61cm).

The sowing instructions are the same as those for A. medeoloides. Germination takes 21–35 days. When the seedlings are large enough to handle, prick them out into 3-inch (8-cm) pots of potting compost. Pot on as necessary.

A. sprengeri

This species of Asparagus is usually grown in hanging-baskets. It has wiry stems and yellow to mid-green foliage. It grows to a height of 12 inches (30cm).

The sowing instructions are the same as those for *A. medeoloides*.

Asperula orientalis
Asperula azurea setosa
ANNUAL WOODRUFF

Hardy annual
Native to the Caucasus

This plant has mid-green leaves and clusters of fragrant pale-blue flowers. It grows to a height of 12 inches (30cm). It thrives on a moist soil in a partially shaded site and is suitable for use as an edging plant.

Sow the seeds March–May in open ground at a depth of $\frac{1}{4}$ inch (6mm). Germination takes 14–21 days. When the seedlings are large enough to handle, thin them out to 4 inches (10cm) apart.

The flowers bloom in July.

Aster, China see *Callistephus chinensis*

Aster, Mexican see *Cosmos bipinnatus*

Aster novi-belgii
MICHAELMAS DAISY

Hardy perennial
Native to the USA

All Asters have daisylike flowers with yellow centres. *A. novi-belgii* has deep-green pointed leaves, and flowers in various colours including mauve-blue, pink, purple, red, and white. Double, semi-double, and dwarf forms are available from seedsmen. The height of the plants varies according to the variety, but for tall varieties is 24–48 inches (61–122cm) and for dwarf varieties is 9–18 inches (23–46cm). Asters grow on any well-drained garden soil and require an open sunny position. Tall varieties will require

Aster novi-belgii 'Ostrich Plume': a double variety.

Aster novi-belgii 'Milady': a dwarf variety.

staking. Dwarf varieties are suitable for rock gardens and for the front of borders.

Sow the seeds March–April at a temperature of 60–68°F (15–20°C). They should be sown at a depth of $\frac{1}{4}$ inch (6mm) in pots or boxes of seed compost. Germination takes 7–10 days. When the seedlings are large enough to handle, prick them out into boxes. Transplant them to their

Aster novi-belgii 'Pepite'.

Aster novi-belgii 'Pinocchio': a dwarf variety.

Aster novi-belgii 'Sinensis': a single variety.

flowering positions during May, spacing them 15 inches (38cm) apart. Alternatively, sow the seeds April–May in the open ground at a depth of $\frac{1}{2}$ inch (1cm). Germination takes 10–14 days. When the seedlings are large enough to handle, thin them out to 15 inches (38cm) apart.

The flowers bloom September–October.

Atriplex hortensis rubra
RED MOUNTAIN SPINACH, RED ORACH

Hardy annual
Native to Europe

This unusual and decorative foliage plant has deep-red leaves and grows to a height of 54 inches (1.37 metres). It forms an excellent ornamental background for the border and may be grown also as a temporary hedge. It requires a well-drained soil.

Sow the seeds in the open ground March–May at a depth of $\frac{1}{4}$ inch (6mm). Germination takes 14–21 days. When the seedlings are large enough to handle, thin them out to 24 inches (60cm) apart.

Aubrieta

Hardy perennial
Native to the mountains of eastern Europe

This evergreen plant has bright flowers in shades of red and purple. It grows to a height of 4 inches (10cm). It needs a sunny position and grows best on a soil which contains lime. It is suitable for use in rockeries.

Sow the seeds April–June in a seed bed at a depth of $\frac{1}{4}$ inch (6mm).

Aubrietia 'Bengal': a semi-double variety.

Germination takes 14–21 days. When the seedlings are large enough to handle, thin them out 12 inches (30cm) apart. Transfer the plants to their permanent positions during September, spacing them 24 inches (60cm) apart.

The flowers bloom March–June.

Auricula see *Primula* × *pubescens*

Azalea Exbury Hybrids

Deciduous shrub

These hybrid shrubs have flowers of yellow, apricot, pink, red, or white. They grow to a height of 48–72 inches (122–183cm). A chalk and lime-free soil is essential for growing Azaleas and they do best in sheltered shady positions.

Sow the seeds February–March on the surface of a sandy peat compost. Keep the temperature at 50–60°F (10–15°C) and do not allow the compost to dry out. Germination takes 21–28 days. When the first true leaves

appear, prick the seedlings out into boxes of compost, spacing them 1 inch (2.5cm) apart. Grow them on until the following April, and then plant them out in a nursery bed. Transfer the shrubs to their permanent positions the following year.

Plants grown from seed will flower 3–4 years after sowing. The flowers bloom May–June.

Baby Blue Eyes see *Nemophila menziesii*

Baby's Breath see *Gypsophila paniculata*

Bachelor's Button see *Gomphrena globosa*

Balm see *Melissa officinalis*

Balm, Bee see *Monarda didyma*

Balsam see *Impatiens balsamina*

Bartonia aurea see *Mentzelia lindleyi*

Basil see *Ocimum*

Basil, Bush see *Ocimum minimum*

Basil, Sweet see *Ocimum basilicum*

Begonia semperflorens

Greenhouse perennial which may be treated as a half-hardy annual
Native to Brazil

The foliage is bright green, although some forms have deep-purple leaves. The flowers range in colour from white, through pink and red, to scarlet. The plant grows to a height of 6–9 inches (15–23cm). Although a greenhouse plant, Begonias flower out of doors during the summer months. They require a position which is well-lit, but not in direct sun, and thrive on a light soil.

As the seeds are very tiny, sow them as thinly as possible on the surface of a

Begonia 'Fiesta': a double-flowered variety.
Begonia Rex is grown for its foliage.

Begonia 'Organdy': a compact F₁ hybrid.

loamless compost and then water them using a watering can with a fine rose. A temperature of 68–78°F (20–25°C) is necessary for germination to take place, and a sheet of glass or polythene may be placed over the seed trays to prevent the compost from drying out. Germination takes 14–28 days. When the seedlings are large enough to handle, prick them out into boxes and grow them on until all risk of late frosts has passed. Then harden them off before planting them out in their flowering positions. Space the plants far enough apart to allow for their spread of up to 9 inches (23cm).

The flowers bloom June–September out of doors, but almost all the year round under glass.

Bellflower see *Campanula*

Bellflower, Chimney see *Campanula pyramidalis*

Bellis perennis
DAISY

Hardy perennial usually grown as a biennial
Native to Europe

The foliage is mid-green and the flowers are white with yellow discs. Cultivated varieties are available in shades of pink and red, and have double flowers. It grows to a height of 4–6 inches (10–15cm). This plant is useful for edging and for rock gardens. Any garden soil is suitable for this plant which grows in either a sunny or shady position.

Sow the seeds May–June in a seed bed at a depth of ¼ inch (6mm). Germination takes 10–14 days. When the seedlings are large enough to handle, thin them out to 3–4 inches (8–10cm) apart. Transfer them to their permanent positions during September, spacing them about 5 inches (13cm) apart.

The flowers bloom April–October.

Bells of Ireland see *Molucella laevis*

Bergamot see *Monarda didyma*

Bird of Paradise Flower see *Strelitzia reginae*

Black-eyed Susan see *Rudbeckia hirta* and *Thunbergia alata*

Blanket Flower see *Gaillardia*

Blazing Star see *Liatris pycnostachya*

Bloodflower see *Asclepias curassavica*

Blue Lace Flower see *Didiscus caeruleus*

Borago officinalis
BORAGE

Hardy annual
Native to Europe

The leaves are covered with silvery hairs. The flowers, which bloom June–September, are blue and star-shaped. The plant grows to a height of 18–36 inches (46–91cm). It thrives on well-drained soil in a sunny position.

Sow the seeds in April in shallow drills where the plants are to grow. Germination takes 14–21 days. When the seedlings are large enough to handle, thin them out to 12 inches (30cm) apart.

Borage leaves may be used fresh in salads and fruit cups.

Borecole, Ornamental see KALE, ORNAMENTAL

Brachycome iberidifolia
SWAN RIVER DAISY

Half-hardy annual
Native to Australia

The flowers resemble small Cinerarias and are blue, pink, mauve, purple, or white in colour. The plant grows to a height of 9 inches (23cm). It requires a rich loamy soil and a sunny sheltered position. It is useful as an edging plant and may be used also at the front of an annual border.

Brachycome iberdifolia.

A double variety of *Bellis perennis.*

Sow the seeds March–May in the flowering site at a depth of $\frac{1}{4}$ inch (6mm). Germination takes 7–10 days. When the seedlings are large enough to handle, thin out to 15 inches (38cm) apart.
The flowers bloom June–September.

Broom, Spanish see *Spartium junceum*

Broom, White see *Cytisus multiflorus*

Browallia

Half-hardy greenhouse perennial
Native to Peru

This plant has flowers of blue, white,
or lavender which resemble Violets in
shape. It grows to a height of 12–24
inches (30–61cm), depending on the
variety. It is usually grown as a pot
plant in the greenhouse or as a
houseplant.

Sow the seeds February–June at a
temperature of 60–68°F (15–20°C) and
a depth of ⅛ inch (3mm). When the
seedlings are large enough to handle,
transplant them to 3-inch (8-cm) pots
containing potting compost.

The flowers bloom June–September.

Some of the many types of cactus that can be
grown from seed.

Browallia 'Jingle Bells'.

Burnet see *Sanguisorba minor*

Burning Bush see *Kochia scoparia*
'*Trichophylla*'

Busy Lizzie see *Impatiens wallerana*

Butterfly Flower see *Schizanthus*
pinnatus

Cabbage, Ornamental

This round-headed cabbage has pink
and white crinkly leaves. It is very
decorative and popular with flower
arrangers.

Sow the seeds March–April at a
temperature of 50–60°F (10–15°C) and
at a depth of ⅛ inch (3mm).
Germination takes 7–10 days. When
the seedlings are large enough to
handle, prick them out into a box and
grow them on. Plant them out in their
permanent positions during June,
spacing them 18 inches (46cm) apart.

The colour of the foliage improves
after the first frost.

Ornamental Cabbage.

Cacalia coccinea see *Emilia flammea*

Cactus

Greenhouse perennial
Native to the Americas

True Cacti come only from the New World, although many cacti-like plants are found in Africa. Seeds are usually supplied as a mixture of varieties by seedsmen. They are grown as pot plants in the greenhouse or as houseplants.

Sow the seeds at any time at a depth of $\frac{1}{8}$ inch (3mm) in a compost consisting of equal parts of loam, grit, and crushed bricks. A temperature of 68–78°F (20–25°C) is necessary for germination to take place. Germination is sporadic, taking from 50 days onwards. Keep the compost moist while germination is taking place. Watering is best done from below. The seedlings grow very slowly and will probably not be ready for potting up individually until about a year after sowing.

Calandrinia umbellata

ROCK PURSLANE

Hardy perennial
Native to Chile

This plant has narrow grey-green leaves and brilliant-magenta flowers. It grows to a height of 6 inches (15cm). It requires a light sandy soil and a sunny position. It is particularly useful as a rockery plant.

Sow the seeds March–April at a temperature of 50–60°F (15–20°C) and at a depth of $\frac{1}{8}$ inch (3mm). Germination takes 10–14 days. When the seedlings are large enough to handle, prick them out into boxes and harden them off. Then plant them out in their flowering positions at the end of May, spacing them 9–12 inches (23–30cm) apart. Alternatively, sow the seeds in their flowering positions during the summer (May–July) at a depth of $\frac{1}{4}$ inch (6mm). When the seedlings are large enough to handle, thin them out to 9–12 inches (23–30cm) apart.

The flowers bloom July–September.

Calceolaria

SLIPPER FLOWER

C. × herbeohybrida

Greenhouse biennial
Hybrids raised from species native to Chile

These hybrids have pouch-shaped flowers in shades of red, yellow, or orange, spotted with crimson. They reach a height of 8–18 inches (20–46cm), depending on the variety. They are grown as pot plants in the greenhouse.

As the seeds are very small, mix them with sand to get an even distribution. Sow the seeds May–July in a shady position at a temperature of 60–68°F (15–20°C). Just press the seeds into the surface of the compost. Germination takes 14–21 days. When

Calceolaria x herbeohybrida 'Perfection'.

Calceolaria x *herbeohybrida* 'Jewel Cluster'.

Calceolaria integrifolia 'Sunshine': an F₁ hybrid.

the seedlings are large enough to handle, prick them out into boxes and grow them on until they are well developed. Then pot them up individually into 3-inch (8-cm) pots of potting compost. Pot on as necessary until the plants are in 6-inch (15-cm) pots.

Bring under cover at the end of September.

The flowers bloom May–July.

C. integrifolia (*C. rugosa*)

Half-hardy perennial best treated as a half-hardy annual
Native to Chile and Peru

The leaves are mid-green and finely wrinkled; the flowers are bright yellow. The plant grows to a height of 8–10 inches (20–25cm). It requires a light soil and grows in either a sunny or partially shaded position. It is a useful summer bedding plant.

Sow the seeds January–March at a temperature of 60–68°F (15–20°C). Just press the seeds into the surface of the compost. Germination takes 14–21 days. When the seedlings are large enough to handle, prick them out into boxes and harden them off. Then plant

them out at the end of May, spacing them 12–15 inches (30–38cm) apart.

The flowers bloom July–September.

Calendula officinalis
POT MARIGOLD, ENGLISH MARIGOLD

Hardy annual
Native to southern Europe

The light green leaves are long and narrow. The flowers are daisylike and usually bright yellow or orange, although pastel shades are available. The plant grows to a height of 24 inches (61cm). Double and dwarf varieties are available from most seedsmen. These plants survive, indeed thrive, on the poorest soil, but do even better on a well-drained soil and in a sunny position.

Sow the seeds March–May at a depth of ½ inch (1cm) where the plants are to flower. Germination takes 10–14 days. When the plants are large enough to handle, thin them out to 12 inches (30cm) apart. Seeds may be sown August–September for early spring flowering.

The flowers bloom from May until the first autumn frosts.

Calendula officinalis 'Orange King' and 'Lemon Queen'.

Calendula officinalis 'Radio' has quilled petals.

Calendula officinalis 'Fiesta Gitana': a double-flowered variety.

shades of pink, red, purple, and white. Both double-flowered and dwarf strains are available. The plant grows to a height of 12–24 inches (30–61cm) but dwarf strains grow only to 8–12 inches (20–30cm). It requires an open sunny position and a well-drained soil.

Sow the seeds during March at a temperature of 61°F (16°C) and at a depth of $\frac{1}{4}$ inch (6mm). When the seedlings are large enough to handle, prick them out into boxes and grow them on. Harden the plants off and transfer them to their flowering positions during May, spacing them 12 inches (30cm) apart.

The flowers bloom from July until the first autumn frosts.

Callistephus chinensis

CHINA ASTER

Half-hardy annual
Native to China

This plant has mid-green toothed leaves and daisylike flowers. Cultivated varieties are available in

Campanula

BELLFLOWER, HAREBELL

Native to Europe

Some species of Campanula are suitable for rockeries and borders, others can be grown as pot plants. All of them require a fertile well-drained soil in order to thrive.

C. carpatica

Hardy perennial

This species has mid-green leaves and bell-shaped flowers in shades of blue, purple, or white. It grows to a height of 6–9 inches (15–23cm) and is suitable for use in borders.

Sow the seeds February–June in a cold frame or greenhouse. Just press the seeds into the surface of the compost. Germination takes 14–21 days. When the seedlings are large enough to handle, prick them out into boxes and grow them on until the following spring. Transfer the plants to their permanent positions during May, spacing them 12–15 inches (30–38cm) apart.

The flowers bloom July–August.

Campanula medium 'Bells of Holland'.

C. medium

CANTERBURY BELL

Hardy biennial

This cottage-garden flower has long hairy leaves and bell-shaped flowers of blue, white, pink, or purple. It grows to a height of 15–36 inches (38–91cm). The most popular form is the cup and saucer variety 'Calycanthema'. There is also the dwarf variety 'Bells of Holland' which grows to a height of 15 inches (38cm). It requires a moderately rich, well-drained soil and a sunny position.

Sow the seeds April–June in a seed bed at a depth of $\frac{1}{4}$ inch (6mm).

Campanula medium: a cup and saucer variety in a mixture of colours.

Campanula medium: an example of a cup and saucer variety.

Germination takes 14–21 days. When the seedlings are large enough to handle, thin them out to 9 inches (23cm) apart. Transplant them to their flowering positions during the autumn, spacing them 12 inches (30cm) apart.

The flowers bloom June–July.

C. persicifolia
PEACH-LEAVED CAMPANULA

Hardy perennial

This species has mid-green leaves and bell-shaped flowers in shades of blue, blue-purple, and white. It grows to a height of 24 inches (61cm) and is suitable for use in borders.

The sowing instructions are the same as those for *C. carpatica*.

The flowers bloom July–August.

C. pyramidalis
CHIMNEY BELLFLOWER

Hardy perennial

This species has heart-shaped leaves and blue or white flowers. It grows to a height of 48 inches (122cm) and may be grown as a pot plant.

The sowing instructions are the same as those for *C. carpatica*, but when the plants are transferred to their permanent positions, space them 18 inches (46cm) apart.

The flowers bloom in May when the plant is grown as a pot plant, and in July when it is grown out of doors.

Campion see *Lychnis* and *Silene schafta*

Canary Creeper see *Tropaeolum peregrinum*

Candytuft see *Iberis umbellata*

Canna
INDIAN SHOT

Half-hardy perennial
Bred from species native to North and South America

This plant has large handsome foliage, which may be green, grey, or dark purple, and spikes of red, yellow, or orange gladiolus-like flowers. It grows to a height of 36–48 inches (91–122cm). It is grown as both a foliage and a flowering plant.

As the seeds are large and very hard, soak them in water for 48 hours. Then chip or slightly file the seed coats before sowing. Sow the seeds January–March at a temperature of 68–78°F (20–25°C) and at a depth of $\frac{1}{2}$ inch (1cm). Since the seeds are so large they may be sown separately in pots or well spaced in deep seed trays. Germination takes 21–28 days. Harden the plants off in May and plant out in June, after all risk of frost has passed, spacing them 12–18 inches (30–46cm) apart. Some plants may flower during their first season. The roots may be lifted and stored like Dahlias in frost-free conditions during the winter. Replant the roots in mid-May at a depth of 6 inches (15cm).

The flowers bloom June–September.

Canterbury Bell see *Campanula medium*

Capsicum annuum
ORNAMENTAL PEPPER

Half-hardy annual
Native to the Tropics

This easily grown pot plant is grown for its bright fruits which set readily. It grows to a height of 7–36 inches (18–91cm), depending on the variety. The variety 'Friesdorfer' is one of the long-stemmed peppers and is useful for cutting. It is widely used as a

Three varieties of *Capsicum annuum*. 'Friesdorfer' (*above*). 'Holiday Cheer' (*above left*). and a foliage variety. 'Variegated Flash' (*left*).

Christmas decoration on the Continent.

Sow the seeds January–July at a temperature of 68–78°F (20–25°C) and at a depth of ¼ inch (6mm). Germination takes 14–21 days. When the seedlings are large enough to handle, prick them out into 3-inch (8-cm) pots of potting compost.

Carnation see *Dianthus caryophyllus*

Carum carvi
CARAWAY

Hardy biennial
Native to Europe

Caraway has feathery leaves and creamy-white flowers. It grows to a height of 15 inches (38cm). It requires a fertile well-drained soil and a position in full sunshine.

Sow the seeds in September where the plants are to flower. When the seedlings are large enough to handle, thin them out to 12 inches (30cm) apart.

The flowers will appear during the second season and the seeds will ripen June–July. However, a dry sunny summer is essential to the ripening process.

Caraway seeds are used in cakes and buns and may be used to flavour cabbage dishes. They are used commercially to flavour the liqueur Kümmel.

Castor-oil Plant see *Ricinus communis* and *Fatsia japonica*

Catananche caerulea

CUPID'S DART

Hardy perennial
Native to Europe

This plant has grey-green leaves and bright-blue cornflowerlike flowers. It grows to a height of 30 inches (76cm). It requires a sunny position on a light well-drained soil. It is suitable for borders and for cutting. It is popular with flower arrangers as it may be dried for use in winter decorations.

Sow the seeds February–March at a temperature of 50–60°F (10–15°C). Germination takes 7–14 days. When the seedlings are large enough to handle, prick them out into boxes and grow them on. Then transfer the plants to their permanent positions in May–June, spacing them 12–15 inches (30–38cm) apart. Alternatively, sow the seeds in open ground April–June at a depth of ¼ inch (6mm). When the seedlings are large enough to handle, thin them out to 12–15 inches (30–38cm) apart.

The flowers bloom June–July.

Catananche caerulea.

Celosia cristata 'Jewel Box'.

Catharanthus roseus see *Vinca rosea*

Catmint see *Nepeta × faassenii*

Celosia argentea cristata

COCKSCOMB

Greenhouse annual
Native to tropical Asia

This is a compact plant with light-green foliage and crested flowerheads of red, yellow, or orange flowers. It grows to a height of 6 inches (15cm). Although this species is usually grown as a pot plant under glass it may be used as a summer bedding plant in the South.

Sow the seeds February–March at a temperature of 60–68°F (15–20°C) and at a depth of ⅛ inch (3mm). Germination takes 7–14 days. When the seedlings are large enough to handle, prick them out into boxes of potting compost and grow them on. Pot them up during May.

The flowers bloom from June onwards, although a check to the plant's growth may cause premature flowering.

C. a. plumosa

C. a. pyramidalis
PRINCE OF WALES' FEATHER

Half-hardy annual
Native to India

This plant has mid-green leaves with a feathery flower plume which is 3–6

Celosia plumosa 'Geisha': a dwarf variety.

Celosia plumosa 'Fairy Fountains'.

out singly into pots and grow them on until they are in 5-inch (13-cm) pots. If the plants are to be used outside, prick them out into trays and harden them off before planting out from late May onwards.

The flowers bloom July–August.

Celsia arcturus

Greenhouse perennial
Native to Crete

This plant has clear-yellow flowers with purple anthers. It grows to a height of 18–24 inches (46–61cm). It is usually grown as a pot plant in a cool greenhouse.

Sow the seeds March–April in seed compost at a depth of $\frac{1}{4}$ inch (6mm) in a heated greenhouse. Keep the temperature at 50–60°F (10–15°C). Germination takes 10–14 days. If you have a cold frame or cool greenhouse the seeds may be sown June–July at a depth of $\frac{1}{2}$ inch (1cm). Germination takes 14–21 days. When the seedlings are large enough to handle, prick them out singly into 3-inch (8-cm) pots containing potting compost.

Cenia barbata see *Cotula barbata*

inches (8–15cm) high. The colours include pink, yellow, crimson, and amber. The plant grows to a height of 12–24 inches (30–61cm), depending on the variety. It is usually grown as a pot plant, but in mild areas it may be used outside in summer bedding schemes, in which case it requires a sheltered sunny position and a rich well-drained soil.

Sow the seeds February–April at a temperature of 68–78°F (20–25°C). The seeds should be sown at a depth of $\frac{1}{8}$ inch (3mm) in a pot or tray containing seed compost. Germination takes 7–10 days. When the seedlings are large enough to handle, prick them

Centaurea

C. cyanus
CORNFLOWER

Hardy annual
Native to Europe

The foliage is grey-green and there are sprays of pink, red, purple, blue or white flowers. The plant grows to a height of 9–36 inches (23–91cm), depending on the variety. Cornflowers thrive on most soils provided that they are given an open sunny position.

Sow the seeds March–May at a depth of $\frac{1}{2}$ inch (1cm) where the plants are to flower. Germination takes 10–14

Centaurea cyanus: a double variety.

Centaurea cyanus: 'Blue Diadem'.

hidden by the foliage. It grows to a height of 18 inches (46cm). Any fertile soil is suitable for this plant but it requires a sunny position. It makes an effective background to bright summer bedding schemes.

Sow the seeds February–March at a temperature of 68–78°F (20–25°C) and at a depth of $\frac{1}{4}$ inch (6mm). When the seedlings are large enough to handle, prick them out into boxes and harden them off. Plant them out in their flowering positions when all danger of frost has passed, spacing them 18–24 inches (46–61cm) apart. Alternatively, sow the seeds in the open ground April–May at a depth of $\frac{1}{2}$ inch (1cm). When the seedlings are large enough to handle, thin them out to 18–24 inches (46–61cm) apart. In both cases germination takes 14–28 days.

The flowers bloom in August.

C. moschata
C. imperialis
SWEET SULTAN

Hardy annual
Native to the Levant

This plant has thin stems and grey-green leaves. The flowers resemble Cornflowers and are white, yellow,

Centaurea imperialis.

days. When the second true leaf appears, thin the seedlings out to 9–15 inches (23–38cm) apart. Sow the seeds August–September to flower the following spring.

The flowers bloom June–September.

C. gymnocarpa

Half-hardy perennial
Native to Sardinia

This species is grown for its silvery-grey foliage. It has small clusters of purple flowers, but these are usually

pink, or purple. The plant grows to a height of 24 inches (61cm). It requires a well-drained soil and sunny position.

Sow the seeds April–May in the open ground at a depth of $\frac{1}{4}$ inch (6mm). Germination takes 7–10 days. When the seedlings are large enough to handle, thin them out to 9 inches (23cm) apart.

The flowers bloom June–September.

Centranthus ruber
Kentranthus ruber
VALERIAN

Hardy perennial
Native to Europe

This plant has large heads consisting of many small red or pink flowers. It grows to a height of 12–15 inches (30–38cm). It requires a sunny position and thrives on poor but well-drained soil. Valerian may be used in borders and as a ground-cover plant. It is often seen naturalized on walls in the western counties of Britain.

Sow the seeds April–June at a depth of $\frac{1}{4}$ inch (6mm) in a nursery bed. Germination takes 14–28 days. When the seedlings are large enough to handle, thin them out 6 inches (15cm) apart and grow them on until the autumn. Transfer the plants to their permanent positions during September, spacing them 9 inches (23cm) apart.

The flowers bloom May–July.

Cerastium tomentosum
SNOW-IN-SUMMER

Hardy perennial
Native to Europe

This species hugs the ground. It has silvery-grey foliage and white star-shaped flowers. It grows to a height of 4–6 inches (10–15cm). It requires a well-drained gritty soil and is an

excellent rockery plant. It may be used also as an edging plant.

Sow the seeds March–May at a depth of $\frac{1}{4}$ inch (6mm) in a sunny position. Germination takes 10–14 days. When the seedlings are large enough to handle, thin them out to 24 inches (61cm) apart. Divide established clumps in March and replant them immediately in their new positions.

The flowers bloom May–July.

Chaenomeles japonica
JAPANESE QUINCE, MAULE'S QUINCE

Hardy shrub
Native to Japan

This shrub has glossy dark-green leaves and orange-scarlet five-petalled flowers. It grows to a height of 36–48 inches (91–122cm). It is a sun-loving plant and thrives on any soil.

Chaenomeles japonica: a hardy shrub.

Sow the seeds November–February in a cool greenhouse or cold frame. Keep them moderately moist. Germination is sporadic, taking from 28 days onwards. When the seedlings are large enough to handle, prick them out singly into 3-inch (8-cm) pots and grow them on. Plant them out in their permanent positions during September, allowing enough space around them to accommodate their spread of about 5 feet (1.5 metres).

The flowers bloom March–May.

Chamomile, Common see *Anthemis nobilis*

Charieis heterophylla see *Kaulfussia amelloides*

Cheiranthus

WALLFLOWER

Hardy perennial usually grown as hardy biennial
Native to Europe

C. × allionii

SIBERIAN WALLFLOWER

The foliage is mid-green and the sweetly-scented flowers, carried on terminal spikes, are orange. It grows to a height of 15 inches (38cm) and

Cheiranthus allionii.

requires a sunny position and a well-drained soil.

Sow the seeds May–June at a depth of ¼ inch (6mm) in a seed bed. Germination takes 10–14 days. Transplant the seedlings to their flowering site during the autumn, spacing them 12 inches (30cm) apart. Alternatively, sow the seeds in the flowering site August–September.

When the seedlings are large enough to handle, thin them out to 12 inches (30cm) apart.

The flowers bloom May–July.

C. cheiri

This is among the most popular of all the late spring-flowering plants. The foliage is dark green and the sweetly-scented flowers, borne on terminal spikes, are available in many shades including red, yellow, orange, white, and rose-pink. The plant grows to a height of 12–15 inches (30–38cm) and dwarf varieties are available which

Cheiranthus cheiri
a giant-flowered variety

grow to a height of 8–12 inches (20–30cm). Wallflowers do well on most soils, but clay or peat soils should be given a dressing of lime before the plants are set out.

Sow the seeds May–June in a seed bed at a depth of ¼ inch (6mm). Germination takes 10–14 days. When the seedlings are large enough to handle, thin them out to 6 inches (15cm) apart. Transplant to the flowering site September–October, spacing tall varieties 12–15 inches (30–38cm) apart, and dwarf varieties 10–12 inches (25–30cm) apart.

The flowers bloom April–June.

Above: Cheiranthus cheiri 'Fire King' and *below:* 'Cloth of Gold'.

A mixture of regular varieties showing the many colours available.

'Suttons Persian Carpet'.

Chelone barbata see *Penstemon barbatus*

Cherry Pie see *Heliotrope*

Cherry, Winter see *Solanum capsicastrum*

Chervil see *Anthriscus cerefolium*

Chinese Lantern see *Physalis alkekengi franchetti*

A dwarf variety of *Cheiranthus cheiri.*

Chives see *Allium schoenoprasum*

Christmas Rose see *Helleborus niger*

Chrysanthemum

C. carinatum
C. tricolor

Hardy annual
Native to the Mediterranean region

This plant has bright-green foliage
and large single flowers. The central
discs of the flowers are purple and the
petals are banded with different
colours. The plant grows to a height of

Chrysanthemum carinatum.

Chrysanthemum coccineum 'Suttons Charm': a
greenhouse variety.

24 inches (61cm) and requires a sunny
position.

Sow the seeds March–May in the
flowering site at a depth of $\frac{1}{4}$ inch
(6mm). Germination takes 10–14 days.
When the seedlings are large enough
to handle, thin them out to 6–9 inches
(15–23cm) apart.

The flowers bloom June–September.

C. coccineum
Pyrethrum roseum

Hardy perennial
Native to the Caucasus

This species has given rise to a number
of varieties. The fernlike foliage is
bright green and the large single, semi-
double, and double flowers resemble
Daisies. They are available in many
different shades of pink and red as well
as white. The plant grows to a height
of 24 inches (61cm). It is usually
grown in the herbaceous border and
requires a sunny position and well-
cultivated soil. It does not tolerate

Chrysanthemum coccineum 'Korean': a border
variety.

drought and must be well watered during dry weather.

Sow the seeds June–July in boxes or pots of seed compost at a depth of $\frac{1}{4}$ inch (6mm). Germination takes 10–21 days. When the seedlings are large enough to handle, prick them out into boxes. Transfer them to the flowering site during the autumn, spacing them 16–18 inches (41–46cm) apart.

The flowers bloom May to July.

C. maximum
SHASTA DAISY

Hardy perennial
Native to Europe

This plant has dark-green foliage and single white flowers with yellow centres. It grows to a height of 18–24 inches (46–61cm) and requires a well-drained soil and sunny position.

Sow the seeds April–July in the flowering site at a depth of $\frac{1}{4}$ inch (6mm). Germination takes 10–14 days. When the seedlings are large enough to handle, thin them out 12–18 inches (30–46cm) apart.

The flowers bloom June–August.

C. parthenium
Matricaria eximia

Hardy perennial usually grown as an annual
Native to Europe

This plant has small double flowers which are usually white or yellow. It

grows to a height of 9–18 inches (23–46cm). It may be used as an edging for borders, as a bedding plant, or as a pot plant. It requires a sunny position and well-drained soil.

Sow the seeds February–March at a temperature of 60–68°F (15–20°C) and at a depth of $\frac{1}{8}$ inch (3mm). Germination takes 10–14 days. When the seedlings are large enough to handle, prick them out into boxes and harden them off. Transfer the plants to their flowering positions at the end of May, spacing them 9–18 inches (23–46cm) apart.

The flowers bloom July–September.

FEVERFEW

This herb is also a member of this species. It has pale-green leaves and daisylike flowers which bloom June–September. It grows to a height of 24 inches (61cm). It requires a well-drained soil and a sunny position.

Sow the seeds March–April in shallow drills in their flowering site. Germination takes 10–21 days. When the seedlings are large enough to handle, thin them out to 10 inches (25cm) apart.

Feverfew is grown as a decorative plant but the dried flowerheads are used commercially in medicinal compounds.

C. ptarmicaeflorum
Pyrethrum ptarmicaeflorum
SILVER LACE

Half-hardy perennial
Native to the Canary Islands

This plant is grown for its finely cut, silver-grey foliage which adds a cool colour to the brightness of summer bedding schemes. It grows to a height of 12 inches (30cm). It requires a sunny position and well-cultivated soil.

Foliage in many different shades from silver to deep green provides a cool corner in this small town garden.

Sow the seeds January–March at a temperature of 60–68°F (15–20°C) and at a depth of ¼ inch (6mm). Germination takes 10–21 days. When the seedlings are large enough to handle, prick them out into boxes and harden them off. Transfer the plants to their permanent positions during May, spacing them 9–18 inches (23–46cm) apart.

Cineraria maritima see *Senecio bicolor*

Clarkia

Hardy annual
Native to western North America

C. elegans

This species has double flowers in white, lavender, purple, salmon-pink, orange, or scarlet which may grow to 2 inches (5cm) in diameter. The plant grows to a height of 24 inches (61cm). Best results are achieved when it is grown in a sunny position on a medium loam.

Clarkia elegans.

Sow the seeds March–June at a depth of ¼ inch (6mm) in the flowering site. Germination takes 10–14 days. When the seedlings are large enough to handle, thin them out 12 inches (30cm) apart. For flowers the following spring, sow the seeds during September in sheltered areas or under cloches.

The flowers bloom July–September.

Clarkia pulchella.

C. pulchella

This species has dainty sprays of flowers in white, crimson, or violet. It grows to a height of 18 inches (46cm).

The sowing instructions are the same as those for *C. elegans*.

The flowers bloom July–September.

Clary see *Salvia sclarea*

Cleome spinosa
SPIDER FLOWER

Half-hardy annual
Native to the West Indies

This plant's thin petals and very long stamens give it a spidery appearance. It grows to a height of 42 inches

Cleome spinosa 'Pink Queen'.

(107cm). It needs a position in full sun and a fertile well-drained soil.

Sow the seeds February–March at a temperature of 60–68°F (15–20°C) and at a depth of $\frac{1}{8}$ inch (3mm). Germination takes 14–28 days. When the seedlings are large enough to handle, prick them out into boxes and harden them off. Plant them out in their flowering positions at the end of May, spacing them 18 inches (46cm) apart. Alternatively, sow the seeds at a depth of $\frac{1}{4}$ inch (6mm) in the flowering site during May. When the seedlings are large enough to handle, thin them out to 18 inches (46cm) apart.

The flowers bloom from late June onwards.

Cobaea scandens
CUP AND SAUCER PLANT

Half-hardy perennial usually grown as a half-hardy annual
Native to Mexico

This plant is a rapid climber with purple-green cup-and-saucer shaped flowers. It grows to a height of 10 feet (3 metres) and quickly covers walls or arches, and also makes a screen. It requires a sunny sheltered position and well-drained soil.

Sow the seeds February–March at a temperature of 60–68°F (15–20°C). Sow the seeds singly on their edges in 3-inch (8-cm) pots containing potting compost at a depth of $\frac{1}{2}$ inch (1cm). Germination takes 14–21 days. Harden the seedlings off and transplant them to their final positions during June, spacing them 24 inches (61cm) apart.

The flowers bloom May–October.

Cocklebur see *Agrimonia eupatoria*

Cockscomb see *Celosia argentea cristata*

Coleus blumei
FLAME NETTLE

Greenhouse perennial
Native to the East Indies

The leaves resemble Nettle leaves and are very ornamental, being brilliantly marked in various colours including green, yellow, orange, red, and maroon. The plant grows to a height of about 18 inches (46cm). It requires plenty of light and sunshine. Remove the flower spikes as soon as they appear.

Coleus blumei 'Rose Wizard'

Coleus blumei 'Rainbow Choice'.

A group of *Coleus blumei* 'Fashion Parade' showing the many different colours of foliage produced by the aptly named 'flame nettle'.

Sow the seeds at any time as long as the temperature is about 60–68°F (15–20°C). They should be sown in pots or boxes of seed compost at a depth of $\frac{1}{8}$ inch (3mm). Germination takes 14–21 days. When the seedlings are large enough to handle, prick them out into small pots and then pot on as necessary until they are in 5-inch (13-

cm) pots. When the plants are about 6 inches (15cm) high, pinch out the growing point in order to encourage a bushy growth.

Columbine see *Aquilegia*

Coneflower see *Rudbeckia hirta*

Consolida ajacis
Delphinium consolida
LARKSPUR

Hardy annual
Native to Europe

This plant has mid-green fernlike leaves with blue, purple, red, pink, or white flowers. It grows to a height of 36–48 inches (91–122cm). It requires a sunny position and thrives on moist soils. There are two popular cultivated strains: the Giant Imperial and the Stock-flowered. Dwarf varieties, which grow to a height of 12 inches (30cm), are also available.

Always sow the seeds in the flowering site as Larkspurs do not like being transplanted. Sow the seeds March–May at a depth of $\frac{1}{4}$ inch

Consolida ajacis 'Suttons Stock-Flowered'.

Consolida ajacis 'Dwarf Rocket': a dwarf variety with spikes of hyacinth-like flowers.

(6mm). Germination takes 14–28 days. When the seedlings are large enough to handle, thin them out 6–9 inches (15–23cm) apart. Seeds sown during September will flower early the following summer.

The flowers bloom June–August.

Convolvulus major see *Pharbitis purpurea*

Convolvulus minor.

Convolvulus tricolor
C. minor

Hardy annual
Native to southern Europe

This plant has dark-green foliage, and trumpet-shaped flowers, each of which has a yellow and white throat. The flowers are available in many colours including blue, red, and pink. The plant grows to a height of 12–15 inches (30–38cm). It grows on any well-drained soil and requires a sunny position.

Sow the seeds March–May at a depth of $\frac{1}{2}$ inch (1cm) where the plants are to flower. Germination takes 10–14 days. When the seedlings are large enough to handle, thin them out to 9–12 inches (23–30cm) apart.

The flowers bloom July–September.

Coral Bells see *Heuchera sanguinea*

Cordyline indivisa
TIE PALM

Half-hardy perennial
Native to New Zealand

This houseplant is grown for its attractive narrow leaves. It grows to a height of 30 inches (76cm) and is very easy to grow from seed.

Sow the seeds at any time at a temperature of 68–78°F (20–25°C) and at a depth of $\frac{1}{8}$ inch (3mm). Germination takes 28–35 days. When the seedlings are large enough to handle, prick them out individually into 3-inch (8-cm) pots of potting compost. Pot on as necessary.

Coreopsis
TICKSEED

Native to North America

C. calliopsis

Hardy annual

A bushy plant which has masses of large flowers in shades of golden-yellow, maroon, or crimson. It grows

Coreopsis calliopsis.

Sow the seeds during March at a temperature of 50–60°F (10–15°C) and at a depth of ¼ inch (6mm). Germination takes 14–21 days. When the seedlings are large enough to handle, prick them out into boxes and grow them on until the autumn. Transfer the plants to their permanent positions during September, spacing them 18 inches (46cm) apart. Alternatively, sow the seeds May–July in a seed bed at a depth of ¼ inch

Coreopsis grandiflora 'Sunburst': a variety with semi-double flowers.

to a height of about 24 inches (61cm). A dwarf form is available which grows to a height of 9 inches (23cm). It requires a sunny position and fertile well-drained soil.

Sow the seeds February–March at a temperature of 50–60°F (10–15°C) and at a depth of ⅛ inch (3mm). Germination takes 14–21 days. When the seedlings are large enough to handle, prick them out into boxes and harden them off. Transfer the plants to their flowering site during May, spacing them 12 inches (30cm) apart. Alternatively, sow the seeds March–May in the flowering site at a depth of ¼ inch (6mm). When the seedlings are large enough to handle, thin them out to 12 inches (30cm) apart.

The flowers bloom July–September.

C. grandiflora

Hardy perennial

This species has narrow, deeply toothed leaves and bright-yellow flowers. It grows to a height of 24 inches (61cm). Like *C. calliopsis*, it requires a sunny position and fertile well-drained soil.

(6mm). When the seedlings are large enough to handle, thin them out to 12 inches (30cm) apart. Transfer the plants to their permanent positions during September, spacing them 18 inches (46cm) apart.

The flowers bloom June–August.

Coriandrum sativum
CORIANDER

Hardy annual
Native to the Near East

The leaves are dark green and the pinky-mauve flowers appear during July. The plant grows to a height of 9 inches (23cm). It grows on any type of soil but requires an open sunny position.

Sow the seeds in shallow drills in April. When the seedlings are large enough to handle, thin them out to 6 inches (15cm) apart.

The seeds may be used to flavour curries and stews and the leaves to flavour soups and broths.

Cornflower see *Centaurea cyanus*

Cosmos bipinnatus
Cosmea
MEXICAN ASTER

Half-hardy annual
Native to Mexico

This plant has fine fernlike foliage with tall flowers which resemble single Dahlias. The colours of the flowers include white, rose-pink, crimson, orange, and yellow. The plant grows to a height of 36 inches (91cm). It requires an open sunny position and should be given plenty of space as it branches vigorously.

Sow the seeds February–March at a temperature of 50–60°F (10–15°C) in boxes or pots of seed compost at a depth of $\frac{1}{4}$ inch (6mm). Germination takes 10–14 days. When the seedlings are large enough to handle, prick them out into boxes of potting compost.

Cosmea bipinnatus 'Sensation': a single-flowered variety.

Transplant them to their flowering site during May, when all danger of frost is past, spacing them 24 inches (61cm) apart. Alternatively, sow the seeds April–May in the flowering site at a depth of $\frac{1}{4}$ inch (6mm). When the seedlings are large enough to handle, thin them out to 24 inches (61cm) apart.

The flowers bloom August–September.

Cotoneaster horizontalis
Deciduous shrub
Native to the Himalayas

The leaves are dark green and the flowers are pink. However, this shrub is usually grown for the profusion of

Cotoneaster horizontalis: a deciduous shrub.

scarlet berries that it bears during the autumn months. It grows to a height of about 24 inches (61cm) and prefers a sunny position and well-drained soil. It makes an excellent decorative covering on low walls and banks.

Sow the seeds December–February under a cold frame. They may be sown either in pots or in a seed bed at a depth of $\frac{1}{2}$–1 inch (1–2.5cm) in a soil-based compost. Germination is sporadic. Transfer the plants to their permanent positions during

September, spacing them far enough apart to allow for their spread of 6–7 feet (1.8–2.1 metres).

The flowers bloom during June.

Cotula barbata
Cenia barbata
PINCUSHION PLANT

Half-hardy annual
Native to South Africa

The narrow leaves of this species are covered with silky hairs. The abundant flowers resemble buttons. The plant has a dwarf tufted shape which gives it the appearance of a pincushion. It grows to a height of 3 inches (8cm). It requires a sunny position and well-drained soil, and is an excellent edging plant.

Sow the seeds February–March at a temperature of 60–68°F (15–20°C) and at a depth of $\frac{1}{8}$ inch (3mm). Germination takes 10–14 days. When the seedlings are large enough to handle, prick them out into boxes and harden them off before planting them out in their flowering positions during May. Alternatively, sow the seeds April–May at a depth of $\frac{1}{4}$ inch (6mm) in the flowering site. When the seedlings are large enough to handle, thin them out to 6 inches (15cm) apart.

The flowers bloom July–August.

Cowslip see *Primula veris*

Cowslip, Giant see *Primula florindae*

Cress, Rock see *Arabis alpina*

Cress, Violet see *Ionopsidium acaule*

Cress, Wall see *Arabis alpina*

Cup and Saucer Plant see *Cobaea scandens*

Cuphea miniata

Half-hardy perennial treated as a half-hardy annual
Native to Mexico

This species has mid-green foliage and bright-red tubular-shaped flowers. It grows to a height of 12 inches (30cm). It is usually grown as a pot plant in a cool greenhouse, but it may also be used as a summer bedding plant.

Sow the seeds March–April at a temperature of 60–68°F (15–20°C) and at a depth of $\frac{1}{4}$ inch (6mm). Germination takes 14–21 days. When the seedlings are large enough to handle, prick them out into boxes and harden them off. Transfer the plants to the flowering site during May, spacing them 18–24 inches (46–61cm) apart. If the plants are to be grown as pot plants, prick out the seedlings when they are large enough to handle into 3-inch (8cm) pots of potting compost. Pot them on as necessary until they are in 6-inch (15-cm) pots.

The flowers bloom June–September.

Cupid's Dart see *Catananche caerulea*

Curry Plant see *Helichrysum angustifolium*

Cyclamen

C. neapolitanum
C. hederifolium

Hardy perennial
Native to southern Europe

This hardy species of Cyclamen has silver marbled foliage and pale-rose flowers. It grows to a height of 4 inches (10cm). It thrives in partially shaded, moist areas of the garden and is particularly suitable for growing in the shelter of trees and shrubs.

Sow the seeds October–March at a depth of $\frac{1}{4}$ inch (6mm). Keep the

Cyclamen persicum 'Puppet': a miniature-flowered variety which does well in small pots of 4 inches (10 cm) diameter

Cyclamen persicum 'Pink Ruffles': a variety which has frilled and ruffled petals.

temperature at 50–60°F (10–15°C). When the seedlings are large enough to handle, prick them out into boxes and grow them on. Transfer the plants to their permanent positions during May of the following year, spacing them 4–6 inches (10–15cm) apart.

The flowers bloom August–November.

C. persicum

Greenhouse perennial
Native to the Levant

The flowers are usually pink, white, or red. The plant grows to a height of about 10 inches (25cm).

Sow the seeds August–March at a temperature of 60–68°F (15–20°C) and at a depth of $\frac{1}{4}$ inch (6mm). Germination takes 21–35 days. When the seedlings are large enough to handle, prick them out individually into 3-inch (8-cm) pots of potting compost. Pot on as necessary. Seeds sown during August will produce flowers at Christmas the following year.

Cynoglossum amabile
HOUND'S TONGUE

Hardy annual
Native to China

This species has grey-green foliage and blue flowers which resemble Forget-me-nots. It grows to a height of 18–24 inches (46–61cm). This plant requires a moderately rich, well-drained soil and a sunny or lightly shaded position.

Sow the seeds March–April in the flowering site at a depth of $\frac{1}{4}$ inch (6mm). Germination takes 10–14 days. When the seedlings are large enough to handle, thin them out to 12 inches (30cm) apart.

The flowers bloom July–August.

Cyperus alternifolius
UMBRELLA PLANT

Greenhouse perennial
Native to Malagasy

This species is grown as a pot plant for its decorative, finely cut, deep-green

Cyperus alternifolius has slender fronds which resemble the spokes of an umbrella.

fronds. It grows to a height of 30 inches (76cm).

Sow the seeds at any time at a temperature of 68–78° (20–25°C) and at a depth of ⅛ inch (3mm). Germination takes 14–28 days. When the seedlings are large enough to handle, prick them out individually into 3-inch (8-cm) pots of potting compost. Repot them as necessary.

Cypress, Summer see *Kochia scoparia 'Trichophylla'*

Cytisus multiflorus
Cytisus albus
WHITE BROOM

Hardy shrub
Native to Spain

This is a bushy species with grey-green foliage. It bears long sprays of small white flowers which are similar in shape to Sweet Peas. It grows to a height of 48 inches (122cm). This species flourishes on most soils, but requires a sunny position.

Before sowing the seeds, soak them in water for 24 hours. Sow 3 seeds in each 3-inch (8-cm) pot of seed compost at a depth of ½ inch (1cm). Germination takes up to 50 days. When the seedlings are large enough to handle, reduce them to 1 per pot. Transfer the plants to their permanent positions during the late summer, spacing them far enough apart to allow for their spread of up to 6 feet (1.8 metres).

The flowers bloom May–June.

Dahlia
Half-hardy perennial
Native to Mexico

Border Dahlias grow to a height of 18–60 inches (46–152cm). The forms available include single- and double-flowered varieties, as well as anemone-flowered, peony-flowered, ball, pompon, cactus, and semi-cactus varieties. The many colours available include cream, white, scarlet, orange, yellow, lavender, and pink. Bedding Dahlias grow to a height of 12–20 inches (30–51cm). They are available in single, semi-double, and double forms and the colours include yellow, pink, and red. Of the many strains of bedding Dahlia available, 'Coltness Hybrids' is one of the best known. All

Dahlia 'Rigoletto': a double-flowered variety.

out into boxes of potting compost. Transplant them to the flowering site during May. Space border Dahlias 18–36 inches (46–91cm) apart, depending on the variety; and space bedding Dahlias 12–24 inches (30–61cm) apart. Border Dahlias may need staking. Tubers may be lifted in the autumn and dried during the winter months for replanting the following season.

The flowers bloom from late July until the first frosts.

Above: Dahlia 'Dandy' : a single-flowered dwarf variety; *below:* 'Quilled Satellite' a double flowered dwarf variety.

Dahlia 'Coltness Hybrids' : a single-flowered dwarf variety.

Dahlias require a sunny or semi-shaded position and well-prepared ground which should be enriched with garden compost or manure.

Sow the seeds February–April at a temperature of 60–68°F (15–20°C) and at a depth of $\frac{1}{4}$ inch (6mm) in pots or boxes of seed compost. Germination takes 10–14 days. When the seedlings are large enough to handle, prick them

Daisy see *Bellis perennis*

Daisy, African see *Arctotis* × *hybrida*

Daisy, Barberton see *Gerbera jamesonii*

Daisy, Kingfisher see *Felicia bergeriana*

Daisy, Livingstone see *Mesembryanthemum criniflorum*

Daisy, Michaelmas see *Aster nova-belgii*

Daisy, Midsummer see *Erigeron speciosus*

Daisy, Shasta see *Chrysanthemum maximum*

Daisy, Swan River see *Brachycome iberidifolia*

Daisy, Transvaal see *Gerbera jamesonii*

Delphinium

Hardy perennial
Native to Europe

The leaves are mid-green and the flowers are borne on spikes which

Delphinium (*above*) 'Suttons Hybridum Strain' which reaches a height of 60 inches (150 cm) and (*below*) 'Blue Fountains' a dwarf hybrid strain with double florets.

grow to a height of 12 inches (30cm) or more. Colours include white, blue, pink, and purple. The plant grows to a height of 3–5 feet (91–152cm), depending on the variety. It requires a well-drained soil which has been deeply dug, and a sunny position.

Sow the seeds May–July in a seed bed at a depth of $\frac{1}{4}$ inch (6mm). Germination takes 14–21 days. When the seedlings are large enough to handle, thin them out to 12 inches (30cm) apart. Transfer the plants to their permanent site during the autumn, spacing them 18–24 inches (46–61cm) apart.

The flowers bloom June–July.

Delphinium consolida see *Consolida ajacis*

Dianthus

PINK, CARNATION

D. alpinus
ROCK PINK

Hardy perennial
Native to Europe

The flowers vary in colour from pink to purple and have a white eye. The plant grows to a height of 6 inches (15cm). This species requires a sunny position and is ideal for use in a rock garden.

Sow the seeds April–July in a cool greenhouse at a depth of $\frac{1}{4}$ inch (6mm). Germination takes 7–10 days. When the seedlings are large enough to handle, prick them out into boxes, spacing them 1 inch (2.5cm) apart. When the plants are well developed, transfer them to their flowering positions, spacing them 6 inches (15cm) apart.

The flowers bloom May–August.

Dianthus barbatus 'Giant Auricula-Eyed' (*above*) and 'Crimson Velvet' (*below*).

'Summer Beauty', an annual variety of *Dianthus barbatus*.

D. barbatus
SWEET WILLIAM

Perennial usually grown as a hardy biennial
Native to Europe

A popular cottage-garden flower which has a sweet scent. The flowerheads consist of densely packed, single or double flowers in various colours including crimson, scarlet, salmon-pink, and pink. The flowers are often marked with other colours including white. The plant grows to a height of 12–24 inches (30–61cm), depending on the variety. There are also dwarf forms which grow to a height of 6–10 inches (15–25cm). Most soils are suitable but it requires a sunny position.

Sow the seeds May–June in a seed bed at a depth of $\frac{1}{4}$ inch (6mm). Germination takes 10–21 days. When the seedlings are large enough to handle, thin them out to 6 inches (15cm) apart. Transplant them to their flowering positions during October, spacing them 8–10 inches (20–25cm) apart.

The flowers bloom June–July.

D. caryophyllus
CARNATION, GILLIFLOWER, CLOVE PINK

Native to Europe

Several perennial, annual, and biennial forms have been derived from this species. Colours vary but include shades of red, pink, and white. The plant grows to a height of 12–18 inches (30–46cm), depending on the variety. It grows best in an open sunny position on well-drained soil.

Perennial varieties
Sow the seeds January–March at a temperature of 60–68°F (15–20°C) and at a depth of $\frac{1}{4}$ inch (6mm). Germination takes 7–10 days. Transplant the seedlings to the flowering positions April–May, spacing them 18 inches (46cm) apart.

A dwarf variety of *Dianthus caryophyllus* (*above*) ; *Dianthus caryophyllus* 'Crimson Knight' (*below*)

Alternatively, sow the seeds in open ground April–July at a depth of $\frac{1}{4}$ inch (6mm). Germination takes 10–14 days. When the seedlings are large enough to handle, thin out to 18 inches (46cm) apart.

The flowers bloom July–September.

Perennial varieties treated as biennials or annuals
Sow the seeds January–March at a temperature of 60–68°F (15–20°C) and at a depth of $\frac{1}{4}$ inch (6mm). Transplant the seedlings to their flowering positions April–May for flowers during August. Plants potted and taken indoors after the first autumn frosts will continue to flower throughout the winter if kept at a temperature of 50–60°F (10–15°C).

D. chinensis
INDIAN PINK

Half-hardy annual
Native to China

This species has flowers similar to those of the English Pink but with looser petals. Both single- and double-flowered strains are available from seedsmen. The plant grows to a height of 12 inches (30cm). It requires a sunny position and well-drained soil.

Sow the seeds January–March at a temperature of 60–68°F (15–20°C) and at a depth of $\frac{1}{8}$ inch (3mm). Germination takes 7–10 days. When the seedlings are large enough to handle, prick them out into boxes and harden them off. Transfer the plants to their flowering positions during May, spacing them 6 inches (15cm) apart. Alternatively, sow the seeds during April at a depth of $\frac{1}{4}$ inch (6mm) in the flowering site. When the seedlings are large enough to handle, thin them out to 6 inches (15cm) apart.

The flowers bloom from July until the first autumn frosts.

D. chinensis 'Heddewigii'
JAPANESE PINK

Half-hardy annual
Bred in Japan

This variety is available in many striking colours including crimson, salmon-pink, scarlet, and white. It grows to a height of 9–12 inches (23–30cm). Single- and double-flowered forms are available from seedsmen. It requires a sunny position and well-drained soil. It is suitable for use in borders and summer bedding schemes.

Sow the seeds January–March at a temperature of 60–68°F (15–20°C) and at a depth of $\frac{1}{8}$ inch (3mm). Germination takes 7–10 days. When the seedlings are large enough to handle, prick them out into boxes and harden them off. Transfer the plants to

Dianthus chinensis 'Heddewigii' 'Queens Court': an F₁ hybrid.
Dianthus chinensis 'Heddewiggii' 'Magic Charms': a dwarf F₁ hybrid.

their flowering positions in late May, spacing them 8–12 inches (20–30cm) apart.

The flowers bloom from early June until the first autumn frosts.

D. deltoides
MAIDEN PINK

Hardy perennial
Native to Europe

This species has dark-green foliage and rosy-crimson flowers. It grows to a height of about 6 inches (15cm). Like *D. alpinus*, it is suitable for use in rock gardens. The sowing instructions are the same as those for *D. alpinus*, but when the plants are transferred to their flowering site, space them 4 inches (10cm) apart.

The flowers bloom from June until the first autumn frosts.

D. plumarius
PINK

Hardy perennial
Native to Europe

This species is available in many colours from seedsmen and is strongly scented. It grows to a height of 9–12 inches (23–30cm).

Sow the seeds March–April at a temperature of 60–68°F (15–20°C) and at a depth of $\frac{1}{4}$ inch (6mm). Germination takes 7–14 days. When the seedlings are large enough to handle, prick them out into boxes and harden them off. Transfer the plants to their flowering positions at the end of May, spacing them 9–12 inches (23–30cm) apart. Seeds sown in a cool greenhouse June–July will flower the following summer.

The flowers bloom from June until the first autumn frosts.

Dianthus chinensis 'Heddewigii' 'Snow Fire': a single-flowered F₁ hybrid.

Didiscus caeruleus
BLUE LACE FLOWER, QUEEN ANNE'S LACE

Half-hardy annual
Native to Western Australia

This species has light-green foliage and lavender-blue flowers which are similar to those of Scabious. It grows to a height of 18–24 inches (46–61cm). It may be grown as a pot plant in a cool greenhouse or planted out of doors in warm sheltered areas. It requires a well-cultivated soil.

Sow the seeds February–March at a temperature of 60–68°F (15–20°C) and at a depth of $\frac{1}{8}$ inch (3mm). Germination takes 10–14 days. When the seedlings are large enough to handle, prick them out individually into 3-inch (8-cm) pots of potting compost. Repot the plants as necessary until they are in 5-inch (13-cm) pots. Plants which are to be transferred to a flowering position in the garden should be planted out in early June and spaced about 9 inches (23cm) apart.

The flowers bloom July–August.

Digitalis purpurea: 'Suttons Excelsior Hybrids'.

Digitalis purpurea
FOXGLOVE

Hardy biennial
Native to Europe

The leaves of this species are deep green and the flowers are borne on spikes which may grow to a height of 36 inches (91cm). The flowers range in colour from red, through pink, to purple. White flowers are also available. Foxgloves grow on most soils, but they prefer them to be slightly acid and do best in partially shaded positions.

As the seeds are very small, sow them May–June in pots of seed compost, barely covering them. Germination takes 14–21 days. When the seedlings are large enough to handle, line them out in a nursery bed, spacing them 6 inches (15cm) apart. Transfer the plants to their flowering positions during September, spacing them 12–24 inches (30–61cm) apart. They will flower the following year.

The flowers bloom June–August.

Dill see *Anethum graveolens*

Dimorphotheca aurantiaca
STAR OF THE VELDT

Hardy annual
Native to South Africa

The flowers resemble Daisies and are white, orange, or pastel coloured. The plant grows to a height of 6–18 inches (15–46cm), depending on the variety. Although this plant succeeds on most soils, it requires a position in full sunshine.

Sow the seeds March–May where the plants are to flower. They should be sown thinly at a depth of $\frac{1}{4}$ inch (6mm). Germination takes 14–21 days. When the seedlings are large enough to handle, thin them out to 12 inches (30cm) apart.

The flowers bloom June–September.

Two varieties of *Dimorphotheca aurantiaca*:
'Suttons New Hybrids' (*above*) and 'Glistening
White' (*below*).

Echinops ritro is an impressive border plant and
can be dried for use in winter floral
arrangements.

Sow the seeds April–June in a seed
bed at a depth of ½ inch (1cm).
Germination takes 14–28 days. When
the seedlings are large enough to
handle, prick them out into a nursery
bed, spacing them 12 inches (30cm)
apart. Transfer the plants to their
permanent positions during the
autumn, spacing them 24 inches
(61cm) apart. The roots of established
plants may be divided at any time
between October and March.

The flowers bloom July–August.

Echinops ritro
GLOBE THISTLE

Hardy perennial
Native to Europe

The leaves are grey-green and the
spherical flowerheads are covered in
small blue flowers. The seedheads are
metallic blue. The plant grows to a
height of 42 inches (107cm). It
requires a well-drained soil and a
sunny position.

Echium lycopsis
Echium plantagineum
VIPER'S BUGLOSS

Hardy annual
Native to Europe

The foliage is mid-green and the blue
or purple flowers are borne on spikes
which grow to a height of 9 inches
(23cm). The plant grows to a height of
3 feet (91cm). Cultivated varieties are
available in white and shades of
mauve, pink, red, and blue. The plant
grows to a height of about 12 inches
(30cm). Most soils are suitable and it

grows in both·sunny and partially
shaded sites.

Sow the seeds March–May in the
flowering site at a depth of ¼ inch
(6mm). Germination takes 14–28 days.
When the seedlings are large enough
to handle, thin them out to 9–12 inches
(23–30cm) apart.

The flowers bloom June–August.

A dwarf variety of *Echium plantagineum*.

Edelweiss see *Leontopodium alpinum*

Emilia flammea
Cacalia coccinea
TASSEL FLOWER

Half-hardy annual
Native to tropical America

This plant has vivid orange-scarlet
flowers which are borne on long stems.
It grows to a height of 12 inches
(30cm). It requires a well-drained soil
and sunny position.

Sow the seeds February–March at a
temperature of 60–68°F (15–20°C) and
at a depth of ⅛ inch (3mm).
Germination takes 14–21 days. When
the seedlings are large enough to
handle, prick them out into boxes and
harden them off. Transfer the plants to
their flowering positions during May,
spacing them 12 inches (30cm) apart.
Alternatively, sow the seeds

Emilia flammea.

April–May at a depth of ¼ inch (6mm)
in the flowering site. When the
seedlings are large enough to handle,
thin them out to 12 inches (30cm)
apart.

The flowers bloom from
midsummer onwards.

Erigeron speciosus
MIDSUMMER DAISY

Hardy perennial
Native to the USA

This species has large daisylike flowers
which are pale mauve with bright-
yellow centres. It grows to a height of
24 inches (61cm). Ideally, it requires a
sunny position and a moist but well-
drained soil. It is excellent for use in
borders and for cutting.

Sow the seeds April–July in a well-
prepared seed bed at a depth of ¼ inch
(6mm). Germination takes 14–21 days.
When the seedlings are large enough
to handle, thin them out to 6 inches
(15cm) apart. Transfer the plants to
their permanent positions during the
autumn, spacing them 12 inches
(30cm) apart. They will flower the
following year.

The flowers bloom June–August.

Eryngium maritimum

SEA HOLLY

Hardy perennial
Native to Europe

This is a striking plant which has silvery-green foliage and amethyst-blue flowerheads which resemble Teazles. It grows to a height of 24 inches (61cm). It requires a sunny position and well-drained soil. This plant is popular with flower arrangers as it may be dried for use in winter decorations.

Sow the seeds April–July in a cold frame or cool greenhouse at a depth of $\frac{1}{4}$ inch (6mm). Germination is sporadic, taking 28–90 days. When the seedlings are large enough to handle, prick them out singly into 3-inch (8-cm) pots of potting compost. Transfer the plants to their permanent positions in the garden at any time from October onwards, spacing them 18 inches (46cm) apart.

The flowers bloom July–September.

Erysimum alpinum

ALPINE WALLFLOWER, FAIRY
WALLFLOWER

Hardy biennial
Native to Europe

This bushy plant is covered with yellow flowers. Varieties with mauve and pale-yellow flowers are also available. It grows to a height of 6–8 inches (15–20cm). It requires a sunny position and a well-drained soil. It thrives on poor soil. It is suitable for use in spring bedding schemes and for the rockery.

Sow the seeds January–February at a temperature of 50–60°F (10–15°C) and at a depth of $\frac{1}{8}$ inch (3mm). Germination takes 10–14 days. When the seedlings are large enough to handle, prick them out into boxes and grow them on. Transfer the plants to their flowering positions during the

Erysimum alpinum 'Golden Gem'.

autumn, spacing them 4–6 inches (10–15cm) apart. Alternatively, sow the seeds May–June in a seed bed at a depth of $\frac{1}{4}$ inch (6mm). Germination takes 14–21 days. When the seedlings are large enough to handle, thin them out to 4 inches (10cm) apart. Transfer the plants to their flowering positions during the autumn, spacing them 4–6 inches (10–15cm) apart.

The flowers bloom during May.

Eschscholzia californica

CALIFORNIAN POPPY

Hardy annual
Native to the western USA

The foliage is blue-green and the plant has a profusion of brightly coloured, poppylike flowers which have delicate silky petals. Colours include rose, orange, yellow, white, and red. The plant grows to a height of 12–15 inches (30–38cm). It succeeds on most soils but does best on sandy ones. It requires a sunny position.

Sow the seeds March–May where the plants are to flower at a depth of $\frac{1}{4}$ inch (6mm). Germination takes 10–14 days. When the seedlings are large enough to handle, thin them out to 9–12 inches (23–30cm) apart. Seeds

Eschscholzia californica 'Ballerina' has fluted petals resembling a ballerina's dress.

sown August–September will flower the following year.

The flowers bloom from mid-June to October.

Eucalyptus globulus
BLUE GUM

Greenhouse tree
Native to Australia

This species has silvery-grey leaves and is usually cultivated as a pot plant

Eurphorbia marginata.

in the greenhouse. It is, however, hardy in areas where the climate is mild. In its native Australia it grows to forest-tree proportions, but when cultivated in a pot it usually grows to about 36 inches (91cm).

Sow the seeds February–April at a temperature of 68–78°F (20–25°C) barely covering them with seed compost. Germination takes 21–28 days. When the seedlings are large enough to handle, prick them out into 3-inch (8-cm) pots of potting compost. Pot on as necessary.

Euphorbia marginata
SNOW ON THE MOUNTAIN, SPURGE

Hardy annual
Native to North America

This species is grown for its very decorative foliage. The leaves are a soft green and become veined with white. It grows to a height of 24 inches (61cm). It grows in either a sunny or partially shaded position.

Sow the seeds March–May at a depth of $\frac{1}{4}$ inch (6mm) in the flowering site. Germination takes 21–28 days. When the seedlings are large enough to handle, thin them out to 12 inches (30cm) apart.

N.B. When cut, the stems exude a milky juice which must not come into contact with the eyes, nose, mouth, or any open wound, as it will set up an intense irritation.

The flowers bloom in September.

Everlasting Flower see Helichrysum

Exacum affine

Greenhouse annual or biennial
Native to India

The foliage is mid-green and the mauve flowers, which resemble Violets, have a fragrance similar to that of Lily of the Valley. The plant

Exacum affine 'Starlight'

grows to a height of 9 inches (23cm). It is grown as a pot plant for use in the greenhouse or as a houseplant.

Sow the seeds February–April at a temperature of 68–78°F (20–25°C) and at a depth of ⅛ inch (3mm). Germination takes 14–21 days. When the seedlings are large enough to handle, prick them out singly into 3-inch (8-cm) pots of potting compost. Pot on as necessary. When this species is grown as a pot plant it will flower about 6 months after sowing. For spring flowers, sow the seeds September–October in a cool greenhouse.

Fair Maids of France see *Ranunculus aconitifolius*

Fatsia japonica
Aralia japonica, Aralia sieboldii
CASTOR-OIL PLANT

Hardy shrub
Native to Japan

This is an evergreen shrub with large glossy leaves and greeny-white flowers. It grows to a height of 9 feet (2.7m). Any soil is suitable for this shrub but it requires a sheltered position. It grows in partial shade as well as in full sun.

Sow the seeds at any time at a temperature of 68–78°F (20–25°C) and at a depth of ½ inch (1cm). Germination takes 21–28 days. When the seedlings are large enough to handle, prick them out singly into 3-inch (8-cm) pots of potting compost. Grow them on, repotting as necessary, until they are in 5-inch (13-cm) pots. Transfer the plants to their permanent positions about a year after sowing.

The flowers bloom October–November.

Felicia bergeriana
KINGFISHER DAISY

Half-hardy annual
Native to South Africa

This species has grey hairy leaves and light blue daisylike flowers. It grows to a height of 6 inches (15cm). It requires a well-drained soil and a sunny, sheltered position. It is used mainly at the front of borders or in rockeries.

Sow the seeds February–April at a temperature of 50–60°F (10–15°C) and at a depth of ⅛ inch (3mm). Germination takes 14–28 days. When the seedlings are large enough to handle, prick them out into boxes and harden them off. Transfer the plants to their flowering positions April–May, spacing them 6 inches (15cm) apart.

The flowers bloom June–September.

Fennel see *Foeniculum vulgare*

Feverfew see *Chrysanthemum parthenium*

Ficus elastica
RUBBER PLANT

Greenhouse tree
Native to tropical Asia

This plant is grown for its attractive dark-green glossy foliage and is a popular houseplant. In its native Asia it is, of course, a forest tree but when grown in the greenhouse as a pot plant it grows to a height of about 48 inches (122cm).

Sow the seeds at any time at a temperature of 68–78°F (20–25°C). Just press the seeds into the surface of the seed compost. Germination is sporadic, taking from 50 days onwards. When the seedlings are large enough to handle, transfer them to 3-inch (8-cm) pots of potting compost. Pot on as necessary.

Flame Nettle see *Coleus blumei*

Flax see *Linum*

Flax, Scarlet see *Linum grandiflorum*

Flower-of-an-hour see *Hibiscus trionum*

Foeniculum vulgare
FENNEL

Hardy perennial
Native to Europe

This plant has bright-green fernlike foliage and tiny yellow flowers which appear July–August. All parts of the plant emit an aniseed smell. It grows to a height of 5 feet (1.5 metres). This plant grows best in a sunny position.

Sow the seeds April–May in shallow drills in open ground. When the seedlings are large enough to handle, thin them out to 18 inches (46cm) apart.

The leaves may be used in salads, in fish dishes, and with vegetables.

Forget-me-not see *Myosotis sylvatica*

Foxglove see *Digitalis purpurea*

Freesia

Half-hardy bulbs bred from South African species

The tubular flowers are very sweetly scented and their colours include cream, white, yellow, mauve, and red. The plant grows to a height of 6–24 inches (15–61cm). In the British Isles it requires some protection, but it thrives out of doors in countries with a Mediterranean climate.

As the seeds are very hard, soak them in water for 24 hours before sowing. They are large enough to be handled individually and are sown at the rate of 10 to a 5-inch (13-cm) pot, 12 to a 6-inch (15-cm) pot, and so on. Sow the seeds at intervals between March and June at a depth of ½ inch (1cm) in potting compost. They require a temperature of 50–60°F (10–15°C) and germination takes 14–21 days. During the summer place the pots outside in a well-lit position, but out of direct sunlight. Bring them back under cover before the first autumn frosts, but keep them in cool conditions, preferably about 50°F (10°C).

The bulbs flower about 9 months

Freesias are sweetly-scented and are available in many beautiful colours. These are from the long-stemmed Van Staaveren Super Strain.

after germinaton, so seeds planted during March will flower in December. By sowing at fortnightly intervals a succession of flowers may be obtained from December to March.

Gaillardia
BLANKET FLOWER
Native to North America

Both half-hardy annual and hardy perennial species have large daisylike flowers which are red and yellow. Both species require a light well-drained soil and sunny position. They are used usually as border plants.

G. aristata

Hardy perennial

This species grows to a height of 30 inches (76cm) although there are dwarf varieties, such as 'Goblin', which grow to about 15 inches (38cm).
    Sow the seeds June–July in a cool greenhouse at a depth of $\frac{1}{4}$ inch (6mm). Germination takes 14–21 days. When

A double-flowered variety of *Gaillardia pulchella*.

A large-flowered variety of *Gaillardia aristata*.

the seedlings are large enough to handle, prick them out into boxes and grow them on. Transfer the plants to their flowering positions during the autumn, spacing them 18 inches (46cm) apart.
    The flowers bloom June–October.

G. pulchella

Half-hardy annual

This species grows to a height of about 24 inches (61cm).
    Sow the seeds February–March at a temperature of 50–60°F (10–15°C) and at a depth of $\frac{1}{4}$ inch (6mm). Germination takes from 14–21 days. When the seedlings are large enough to handle, prick them out into boxes and harden them off. Transfer the plants to their flowering positions during May, spacing them 12 inches (30cm) apart.
    The flowers bloom July–October.

Gaura lindheimeri

Hardy perennial
Native to the southern USA

The flowers are white, flushed with pink. The plant grows to a height of

18–24 inches (46–61cm). It may be grown as a border plant and is useful as a cut flower. It requires a well-drained soil and sunny position.

Sow the seeds March–April at a depth of ½ inch (1cm) in the flowering site. Germination takes 21–28 days. When the seedlings are large enough to handle, thin them out to 12 inches (30cm) apart.

The flowers bloom August–September.

Gayfeather see *Liatris pycnostachya*

Gazania x hybrida

Half-hardy annual
Native to South Africa

Gazania x *hybrida* must not be planted in the shade.

This species has dark-green foliage and large daisylike flowers, usually in shades of orange. It grows to a height of 9 inches (23cm). It requires a sunny position and well-drained soil.

Sow the seeds February–March at a temperature of 60–68°F (15–20°C) and at a depth of ¼ inch (6mm). Germination takes 14–21 days. When the seedlings are large enough to handle, prick them out into boxes and harden them off. Transfer the plants to

their flowering positions at the end of May or early in June when all danger of frost has passed, spacing them 12 inches (30cm) apart.

The flowers bloom July–September.

Gentiana

G. acaulis
TRUMPET GENTIAN, GENTIANELLA

Hardy perennial
Native to Europe

This species has trumpet-shaped flowers of brilliant blue. It grows to a height of 4 inches (10cm). It requires a sunny position and leafy soil.

Germination of Gentians is likely to be erratic. In order to overcome this, sow the seeds in boxes containing light porous soil, giving good drainage. During October sink the containers outdoors as freezing helps germination. Leave the containers exposed to frost and snow, but protect them from heavy rain. In spring transfer the seedlings to 3-inch (8-cm) pots of potting compost and grow them on in a cool greenhouse until the autumn. Transfer the plants to their permanent positions during September, spacing them about 18 inches (46cm) apart.

The flowers bloom March–June.

G. septemfida

Hardy perennial
Native to Asia Minor

This species has mid-green foliage and deep-blue flowers. It grows to a height of 9–12 inches (23–30cm). This is considered to be the easiest Gentian to grow. It requires the same soil conditions as *G. acaulis* and the sowing instructions are the same.

The flowers bloom July–August.

A mature rock garden containing ageratum, alyssum and phacelia. The conifers add height and year-round colour.

Geranium see *Pelargonium*

Gerbera jamesonii

BARBERTON DAISY, TRANSVAAL DAISY

Half-hardy perennial
Native to South Africa

This species has hairy leaves and orange-scarlet, long-petalled flowers. Hybrid varieties have flowers in many pastel shades. The plant grows to a height of 18 inches (46cm). Gerberas may be grown as pot plants and as summer bedding plants. They require a sunny position during the summer and the protection of a cool greenhouse during the winter.

Gerbera jamesonii.

Sow the seeds February–March at a temperature of 50–60°F (10–15°C). Just press the seeds into the surface of the seed compost. Germination takes 14–21 days. When the seedlings are large enough to handle, prick out those that are to be grown as pot plants into 3-inch (8-cm) pots of potting compost. Pot on as necessary. Prick out into boxes those that are to be used in summer bedding schemes and harden them off. Transfer them to their flowering positions during May, spacing them about 24 inches (61cm) apart.

The flowers bloom May–August.

Geum chiloense

Hardy perennial
Native to Chile

This plant has mid-green leaves and saucer-shaped flowers of red, yellow, or orange. It grows to a height of 18–24 inches (46–61cm). Most soils are suitable for Geum and it thrives in either a sunny or partially shaded position.

Sow the seeds January–February at a temperature of 60–68°F (15–20°C) and at a depth of ⅛ inch (3mm). Germination takes 21–28 days. When the seedlings are large enough to handle, prick them out into boxes of potting compost and grow them on. Transfer the plants to their flowering positions during September, spacing them 12–18 inches (30–46cm) apart. Alternatively, sow the seeds April–July in a cold frame or cool greenhouse at a depth of ¼ inch (6mm).

The flowers bloom July–September.

Geum chiloense 'Mrs Bradshaw' (scarlet) and 'Lady Stratheden' (golden yellow).

Gilia lutea

Leptosiphon hybridus
STARDUST

Hardy annual
Native to California

This species has mid-green lobed leaves and brightly coloured, star-

87

shaped flowers. It grows to a height of 4–6 inches (10–15cm). It requires a light well-drained soil and sunny position. It is suitable for use in the rockery and in the annual border.

Sow the seeds March–May at a depth of $\frac{1}{4}$ inch (6mm) in the flowering site. Germination takes 14–21 days. When the seedlings are large enough to handle, thin them out to 4 inches (10cm) apart.

The flowers bloom June–September.

Gilliflower see *Dianthus caryophyllus*

Globe Amaranth see *Gomphrena globosa*

Gloxinia see *Sinningia speciosa*

Godetia

Hardy annual
Native to the western USA

A compact plant with mid-green foliage. The flowers may be single, double, or semi-double. Single flowers resemble Poppies, while double and semi-double varieties have frilled

petals. Many colours are available including crimson, pink, white, cherry-red, and salmon-pink. The plant grows to a height of 12–15 inches (30–38cm). Dwarf varieties are also available and these grow to a height of 9–10 inches (23–25cm). Godetia is a very useful border plant which requires a position in full sunlight. It grows on most soils.

Sow the seeds March–April at a depth of $\frac{1}{4}$ inch (6mm) where the plants are to flower. Germination takes 10–14 days. When the seedlings are large enough to handle, thin them out to 6 inches (15cm) apart. Seeds sown during September will flower early the following summer.

The flowers bloom June–August.

Gomphrena globosa

BACHELOR'S BUTTON, GLOBE AMARANTH

Half-hardy annual
Native to India

This bushy species has light-green leaves and white, red, or purple flowers. It grows to a height of 12 inches (30cm). It requires a sunny position and well-drained soil. It is a

A dwarf variety of *Godetia*.

Gomphrena globosa is an 'everlasting' flower.

popular flower with flower arrangers as it may be dried for use in winter decorations.

Sow the seeds March–April at a temperature of 68–78°F (20–25°C) and at a depth of $\frac{1}{4}$ inch (6mm). Germination takes 14–28 days. When the seedlings are large enough to handle, prick them out into boxes and harden them off. Transfer the plants to their flowering positions during May, spacing them 6–9 inches (15–23cm) apart.

The flowers bloom July–September.

Gourd, Ornamental

Half-hardy annual
Native to tropical America

This non-edible curcubit is grown for its small decorative fruits which have a wide variety of shapes and colours. The flowers are yellow. It rambles over the ground or climbs a trellis, and requires a well-drained soil and a position in full sunshine.

Sow the seeds in late May where the plants are to grow at a depth of $\frac{1}{4}$ inch (6mm). Space the plants 36 inches (91cm) apart.

Allow the fruits to ripen fully before they are picked and dried.

The flowers bloom July–September.

Ornamental gourds grown in many different shapes and colours.

Ornamental grasses are much favoured by flower arrangers for winter decorations.

Grasses (Ornamental)

Hardy annual

Seeds of ornamental grasses are usually sold by seedsmen in packets of mixed varieties. They are grown as decorative plants and are useful to flower arrangers as they may be dried and used in winter decorations. Most cultivated varieties grow to a height of 12–24 inches (30–61cm). Any ordinary garden soil is suitable and a sunny position is preferred.

Sow the seeds March–May in the open ground at a depth of $\frac{1}{4}$ inch (6mm). Germination takes 14–21 days. When the seedlings are large enough to handle, thin them out to 9–12 inches (23–30cm) apart.

If the grasses are to be dried, cut them before the flowerheads are fully mature.

Grevillea robusta

Half-hardy tree
Native to Australia

This tree has fernlike foliage. It prefers an acid soil and requires a sunny sheltered position in the garden. However, it is usually grown as a pot plant in a cool greenhouse. When grown as a pot plant, it reaches a height of 3–6 feet (91–183cm).

Sow the seeds February–April at a depth of ¼ inch (6mm) in a compost made up of equal parts (by volume) of sand, peat, and loam. The temperature should be 68–78°F (20–25°C).

Grevillea robusta is grown as a greenhouse shrub in the UK.

Germination takes 14–28 days. When the seedlings are large enough to handle, prick them out singly into 3-inch (8-cm) pots. Pot on as necessary.

Gum, Blue see *Eucalyptus globulus*

Gypsophila

G. elegans

Hardy annual
Native to the Levant

This plant has narrow leaves and numerous white, pink, or rose-coloured flowers. It grows to a height of 24 inches (61cm). It requires a well-drained soil and sunny position.

Sow the seeds March–May at a depth of ¼ inch (6mm) where the plants are to flower. Germination takes 10–14 days. When the seedlings are large enough to handle, thin them out

to 12 inches (30cm) apart.

The flowers bloom May–September.

G. paniculata
BABY'S BREATH

Hardy perennial
Native to Russia

This variety has grey-green leaves and white or pink flowers. It grows to a height of 36 inches (91cm). It requires a well-drained soil and a sunny position.

Sow the seeds February–April at a temperature of 50–60°F (10–15°C). The seeds should be sown in pots or trays containing seed compost at a depth of ¼ inch (6mm). Germination takes 14–21 days. Transfer the plants May–June to their flowering positions, spacing them 24–36 inches (61–91cm) apart.

The flowers bloom June–August.

Harebell see *Campanula*

Helianthemum
ROCK CISTUS, SUN ROSE

Hardy perennial
Native to Europe

This species has grey foliage and buttercup-shaped flowers in many bright colours including orange, pink, and scarlet. It grows to a height of 10 inches (25cm). It requires a warm, sunny position and a well-drained soil. It is suitable for use in rockeries or on dry walls.

Sow the seeds March–May at a temperature of 50–60°F (10–15°C) and at a depth of ¼ inch (6mm). Germination takes 14–28 days. When the seedlings are large enough to handle, prick them out into boxes and grow them on until the autumn. Transfer the plants to their flowering positions during September, spacing them 24 inches (61cm) apart.

The flowers bloom May–July.

Two varieties of *Helianthus annuus* 'Sunburst' (*above*) and 'Large-Flowered Hybrids' (*below*). A very popular garden plant which adds height to the border planting scheme.

Helianthus annuus
SUNFLOWER

Hardy annual
Native to the USA

This plant has mid-green toothed leaves and enormous daisylike flowers which may grow to 12 inches (30cm)

or more in diameter. The flowers of cultivated varieties are available in many colours ranging from pale primrose to copper-bronze. The central discs are brown or purple. Sunflowers grow to a height of 3–10 feet (91–300cm). They require a sunny position and well-drained soil. The taller varieties require staking.

Sow the seeds March–May at a depth of ½ inch (1cm) in the flowering site. When the seedlings are large enough to handle, thin them out to 12–18 inches (30–46cm) apart.

The flowers bloom July–September.

Helichrysum
EVERLASTING FLOWER, STRAW FLOWER

H. angustifolium
CURRY PLANT

Half-hardy perennial
Native to the Mediterranean region

The silver-grey needle-shaped leaves are covered with down and the flowers are small and yellow. The plant grows to a height of 8–15 inches (20–38cm). It requires a sunny position and well-drained soil.

Sow the seeds February–March at a temperature of 60–68°F (15–20°C). The seeds should be sown at a depth of ¼ inch (6mm) in pots or boxes containing seed compost. When the seedlings are large enough to handle, thin them out and then transfer them to their permanent positions during the early summer, spacing them 24 inches (61cm) apart.

The flowers bloom June–August.

The leaves may be used to garnish meat dishes and curries as they have an aroma of curry. They should be used fresh as the aroma fades when they are dried.

Below: 'Bright Bikini'. All varieties of this *Helichrysum bracteatum* can be dried for use in winter floral arrangements.

A tall variety of *Helichrysum bracteatum* which grows to a height of 36 inches (91 cm).

A dwarf variety of *Helichrysum bracteatum*: 'Hot Bikini'.

H. bracteatum
H. macranthum

Half-hardy perennial
Native to Australia

This variety has mid-green leaves and bright daisylike flowers of orange, red, pink, yellow, or white. It grows to a height of 36 inches (91cm). It requires a light well-drained soil and sunny position. It is a very useful plant for flower arrangers as the flowers may be dried and used in winter floral decorations.

Sow the seeds March–May in open ground at a depth of ¼ inch (6mm). Germination takes 10–14 days. When the seedlings are large enough to handle, thin them out to 12 inches (30cm) apart.

The flowers bloom July–September.

Heliotrope
CHERRY PIE

Half-hardy perennial usually treated as a half-hardy annual
Native to South America

This plant has dark-green foliage and tightly packed heads of flowers, which range in colour from dark violet,

Heliotrope 'Marine'

through lavender, to white. It grows to a height of 15 inches (38cm). It requires a sunny position and well-drained soil.

Sow the seeds February–March at a temperature of 60–68°F (15–20°C) and at a depth of $\frac{1}{4}$ inch (6mm). Germination takes 14–28 days. When the seedlings are large enough to handle, prick them out into boxes and harden them off. Transfer the plants to their flowering positions during late May, spacing them 12–15 inches (30–38cm) apart.

The flowers bloom June–October.

Helipterum

H. manglesii
Rhodanthe manglesii

Hardy annual
Native to Western Australia

This plant has single red or white daisylike flowers with yellow centres. It grows to a height of 12 inches (30cm). It grows on poor but well-drained soil and requires a sunny position. It is suitable for planting in the border or the rockery. It is popular with flower arrangers as it may be dried and used in winter decorations.

Sow the seeds April–June in the flowering site at a depth of $\frac{1}{2}$ inch (1cm). Germination takes 14–28 days.

Helipterum roseum is an 'everlasting' flower with a papery texture. Double and semi-double forms are available.

When the seedlings are large enough to handle, thin them out to 6 inches (15cm) apart.

The flowers bloom July–September.

H. roseum
Acroclinium roseum

Hardy annual
Native to Australia

The grey-green leaves are narrow and pointed. The daisylike flowers, which are white or pink, have a papery texture and are excellent for drying. The plant grows to a height of 15 inches (38cm). It grows on poor but well-drained soil and requires a sunny position.

Sow the seeds during March at a temperature of 50–60°F (10–15°C) and at a depth of $\frac{1}{4}$ inch (6mm). Germination takes 14–21 days. When the seedlings are large enough to handle, prick them out into boxes and harden them off. Transfer the plants to their flowering positions during May, spacing them 6 inches (15cm) apart. However, as Helipterum does not transplant well it is more satisfactory to sow the seeds in the flowering site April–May. Sow the seeds at a depth of $\frac{1}{4}$ inch (6mm). When the seedlings are large enough to handle, thin them out to 6 inches (15cm) apart.

The flowers bloom July–August.

Helleborus niger
CHRISTMAS ROSE

Hardy perennial
Native to Europe

This plant has evergreen foliage and large saucer-shaped, five-petalled, white-tinged flowers with a mass of golden-yellow anthers. It grows to a height of 9–12 inches (23–30cm). It grows best on well-drained moist soil and in a shady position.

Sow the seeds in a cold frame at a
depth of ½ inch (1cm) during the
autumn so that they are exposed to the
frost. They should germinate during
April but are sometimes very slow.
When the seedlings are large enough
to handle, prick them out into a
nursery bed, spacing them 9 inches
(23cm) apart. Transfer the plants to
their permanent positions during the
following autumn, spacing them 18
inches (46cm) apart. The plants will
flower for the first time when they are
2–3 years old. Protect the blooms with
cloches.

The flowers bloom December–
March.

Hesperis matronalis
SWEET ROCKET

Hardy perennial
Native to Europe

This is a favourite cottage-garden
flower. It has dark-green foliage and
pyramid-shaped spikes of small,
sweetly scented flowers in shades of
mauve, purple, or white. It grows to a
height of 24 inches (61cm). This plant
grows in full sun or partial shade and
may be naturalized in woodland or
shrubberies.

For flowers the following year, sow
the seeds April–June at a depth of
¼ inch (6mm) in a well-prepared seed
bed. Germination takes 7–10 days.
When the seedlings are large enough
to handle, transplant them to a nursery
bed, spacing them about 9 inches
(23cm) apart. Transfer the plants to
their permanent positions during the
autumn, spacing them 18 inches
(46cm) apart.

The flowers bloom during June.

Heuchera sanguinea
CORAL BELLS

Hardy perennial
Native to North America

The evergreen foliage has rounded

leaves and bears graceful sprays of
vivid coral-red flowers. The plant
grows to a height of 12–18 inches
(30–46cm). It requires a light well-
drained soil. It is useful everywhere in
the garden and even thrives in partial
shade.

Sow the seeds March–April in a cold
frame or greenhouse at a depth of
⅛ inch (3mm). Germination takes 14–28
days. When the seedlings are large
enough to handle, prick them out into
boxes and grow them on. Transplant
them to a nursery bed during June,
spacing them about 12 inches (30cm)
apart. Transfer the plants to their
permanent positions during October,
spacing them 18 inches (46cm) apart.

The flowers bloom June–September.

Hibiscus moscheutos.

Hibiscus

H. moscheutos

Hardy perennial
Native to the USA

The exotic blooms have a diameter of
up to 11 inches (28cm). The flowers
are white, pink, rose, or crimson. The
plant grows to a height of 5 feet (1.5
metres). This species has proved to be
hardy in Britain and over-winters on
well-drained soil.

Sow the seeds January–February at a temperature of 68–78°F (20–25°C) and at a depth of ⅛ inch (3mm). Germination takes 14–28 days. When the seedlings are large enough to handle, prick them out into boxes and harden them off. Transfer the plants to their permanent positions during May, spacing them 24 inches (61cm) apart.

These plants will flower during their first year in favourable conditions.

The flowers bloom August–September.

H. trionum
FLOWER-OF-AN-HOUR

Hardy annual
Native to the Old World tropics

The primrose-coloured flowers have a maroon eye. The plant grows to a height of 15 inches (38cm). It grows on any well-drained soil but requires a sunny position.

Sow the seeds March–May at a depth of ¼ inch (6mm) in the flowering site. Germination takes 14–28 days. When the seedlings are large enough to handle, thin them out to 12 inches (30cm) apart.

The flowers bloom August–September.

Hibiscus trionum.

Holly, Sea see *Eryngium maritimum*

Hollyhock see *Althaea rosea*

Honesty see *Lunaria annua*

Horehound, White see *Marrubium vulgare*

Hound's Tongue see *Cynoglossum amabile*

Hypericum calycinum
ROSE OF SHARON, ST JOHN'S WORT

Hardy perennial
Native to the Levant

A low-growing bushy species which provides excellent ground cover. It has

Hypericum calycinum.

mid-green foliage and large bright-yellow flowers. It grows to a height of 12–18 inches (30–46cm). It grows best in a sunny position and well-drained soil, but tolerates shade and may, therefore, be planted under trees.

Sow the seeds February–April at a depth of ¼ inch (6mm) in a soil-based compost in a cool greenhouse. Germination takes 14–28 days. When the seedlings are large enough to handle, prick them out singly into 3-inch (8-cm) pots of potting compost and grow them on. Transfer the plants to their permanent positions during the following April, spacing them well apart as they spread rapidly.

The flowers bloom June–September.

Hypoestes sanguinolenta
POLKA DOT PLANT

Greenhouse perennial
Native to Malagasy

This species is grown as a pot plant for its attractive foliage, which is bright green splashed with pink dots. The plant grows to a height of 12–18 inches (30–46cm). It is suitable for growing in the greenhouse and may be used as a houseplant.

Sow the seeds at any time at a temperature of 60–68°F (15–20°C) and at a depth of ¼ inch (6mm). Plant the seeds singly in 5-inch (13-cm) pots of potting compost.

Hypoestes sanguinolenta is one of the easiest foliage plants to grow from seed.

Hyssopus officinalis
HYSSOP

Hardy perennial
Native to the Mediterranean region and central Asia

This species has mid-green leaves and purplish-blue tube-shaped flowers which appear from July to September. The plant grows to a height of 24 inches (61cm). There are also white and pink forms. It requires a light well-drained soil.

Sow the seeds in shallow drills at a depth of ¼ inch (6mm) in open ground during April. Germination takes 14–21 days. When the seedlings are large enough to handle, thin them out to 12 inches (30cm) apart. In order to make a hedge, space the plants 9–12 inches (23–30cm) apart.

The young leaves have a bitter minty taste and may be used fresh in salads. The leaves may be used also either fresh or dried to flavour soups and stuffings.

Iberis umbellata
CANDYTUFT

Hardy annual
Native to Europe

This plant has narrow leaves and clusters of white, red, or purple flowers. It grows to a height of 6–15 inches (15–38cm), depending on the variety. The plant thrives on poor soil but ideally requires a sunny position and well-drained soil. It is suitable for use in edging borders.

A dwarf variety of *Iberis umbellata*: 'Dwarf Fairy Mixed'.

Sow the seeds March–May in the flowering site at a depth of $\frac{1}{4}$ inch (6mm). Germination takes 10–14 days. When the seedlings are large enough to handle, thin them out to 9 inches (23cm) apart.

If successive sowings are made, flowers will bloom June–September.

Impatiens

I. balsamina
BALSAM, TOUCH-ME-NOT

Half-hardy annual
Native to Asia

This species has pale-green leaves and pink, white, crimson, or purple flowers. It grows to a height of 30 inches (76cm). There is a double-flowered form, 'Camellia Flowered', which grows to a height of 18 inches (46cm), and a dwarf double-flowered form, 'Dwarf Tom Thumb', which grows to a height of 8 inches (20cm). This plant is often grown indoors as a pot plant. The tall variety is suitable for use outside and survives both dry and stormy weather. Any well-drained soil is suitable for this species and it grows in either a sunny or partially shaded position.

Sow the seeds February–April at a temperature of 60–68°F (15–20°C) and at a depth of $\frac{1}{4}$ inch (6mm). Germination takes 10–14 days. When the seedlings are large enough to handle, prick out singly into 3-inch (8-cm) pots of potting compost those that are to be used as pot plants and prick out into boxes and harden off those that are to be planted outside. Transfer the outdoor plants to their flowering positions at the end of May, spacing them 18 inches (46cm) apart.

The flowers bloom June–September.

I. wallerana
Impatiens sultani
BUSY LIZZIE

Greenhouse perennial usually grown as a half-hardy annual
Native to East Africa

The plant is covered with a profusion of red, purple, orange, or pink flowers which are flat and five-petalled. The leaves are bright green. The F_1 hybrids have large flowers which are sometimes striped with white. The

Impatiens wallerana 'Super Elfin Mixed'

Impatiens wallerana 'Futura'

plant grows to a height of 8–9 inches
(20–23cm). This species is very
popular as a houseplant and may be
planted outdoors as part of a summer
bedding scheme. It grows on any
ordinary well-drained garden soil.

Sow the seeds March–April at a
temperature of 60–68°F (15–20°C) and
at a depth of $\frac{1}{8}$ inch (3mm).
Germination takes 14–21 days. When
the seedlings are large enough to
handle, prick them out into boxes.
Transplant them to the flowering site
in June, spacing them about 12 inches
(30cm) apart.

The flowers bloom April–
September.

Incarvillea delavayi

Hardy perennial
Native to China

This plant has rosy-purple trumpet-
shaped flowers which develop before
its mid-green leaves are fully
expanded. It grows to a height of 8
inches (20cm). It requires a well-
drained soil and sunny position and is
suitable for growing at the front of the
perennial border.

Sow the seeds March–April at a
depth of $\frac{1}{4}$ inch (6mm) in a nursery

bed. Germination takes 14–28 days.
When the seedlings are large enough
to handle, thin them out to 15 inches
(38cm) apart. Transfer them to their
permanent positions during the
following spring.

The flowers bloom May–July.

Indian Shot see *Canna*

Ionopsidium acaule
VIOLET CRESS

Hardy annual
Native to Portugal

This low-growing plant has small
four-petalled flowers which are pale
mauve or white tinged with purple. It
grows to a height of 2–3 inches
(5–8cm). It requires moist soil and
partial shade. It is suitable for the
rockery and may be used as a carpeting
plant in borders.

Sow the seeds March–May at a
depth of $\frac{1}{4}$ inch (6mm) in the flowering
site. Germination takes 14–28 days. It
is unnecessary to thin out the
seedlings.

The flowers bloom June–September.

Ipomoea rubro-caeruleum see
Pharbitis tricolor

Jacaranda ovalifolia
MIMOSA-LEAVED EBONY TREE

Greenhouse tree
Native to Brazil

This shrub has fernlike leaves. It
grows to a height of 10 feet (3 metres).
It is usually grown as a decorative
plant in a conservatory or greenhouse,
but it may be used as a summer

Incarvillea delavayi.

bedding plant in the garden where it adds height to the bedding scheme. Young plants may be used as decorative pot plants in the house.

Sow the seeds February–June at a temperature of 68–78°F (20–25°C). The seeds should be planted in seed compost on their edges and at a depth of $\frac{1}{2}$ inch (1cm). Germination takes 14–28 days. When the seedlings are large enough to handle, transplant them singly into 3-inch (8-cm) pots of potting compost. Pot on as necessary.

Jacobea see *Senecio elegans*

Jerusalem Cross see *Lychnis chalcedonica*

Joseph's Coat see *Amaranthus tricolor*

Kalanchoe blossfeldiana

Greenhouse perennial
Native to Malagasy

This species has mid-green fleshy leaves and soft reddish-scarlet flowers. It grows to a height of 12 inches (30cm). It is grown both as a greenhouse plant and as a decorative houseplant. Water it well during the summer, and keep it in a warm

Kalanchoe blossfeldiana 'Dwarf Vulcan

position and only slightly moist during the winter.

For winter-flowering plants, sow the seeds March–April at a temperature of 60–68°F (15–20°C). Plant the seeds $\frac{1}{8}$ inch (3mm) deep in trays or boxes of seed compost. Germination takes 14–28 days. When the seedlings are large enough to handle, prick them out singly into 3-inch (8-cm) pots of potting compost. Pot on as necessary.

Plants grown from seeds sown March–April will flower February–May.

Kale, Ornamental

ORNAMENTAL BORECOLE

Hardy biennial

This plant is grown in the flower garden as a decorative foliage plant. The seeds are usually supplied as a mixture rather than as a single colour. The frilled foliage is available in many shades of purple and purplish-green. It requires a well-drained and well-cultivated soil.

Sow the seeds March–May in a seed bed at a depth of $\frac{1}{4}$ inch (6mm). Germination takes 7–10 days. When the seedlings are large enough to handle, thin them out to 9 inches (23cm) apart. Transplant them to their final positions during July, spacing them 15–24 inches (38–61cm) apart.

Kaulfussia amelloides
Charieis heterophylla

Hardy annual
Native to South Africa

This species has tiny, bright-blue, daisylike flowers. It grows to a height of 6 inches (15cm). It flourishes in a sunny part of the garden and may be used for edging borders or in a rockery.

Sow the seeds March–May at a

depth of ¼ inch (6mm) in the flowering site. Germination takes 7–14 days. When the seedlings are large enough to handle, thin them out to 4–6 inches (10–15cm) apart.

The flowers bloom from late June onwards.

Kentranthus ruber see *Centranthus ruber*

Kniphofia
RED HOT POKER, TORCH LILY, TRITOMA

Hardy perennial
Native to South Africa

This plant has large spikes of brightly coloured flowers. The many colours available include red, yellow, orange, and salmon-pink. It grows to a height of 36 inches (91cm). It thrives on any well-drained garden soil, but requires a position in full sun.

Sow the seeds January–April at a temperature of 60–68°F (15–20°C) and at a depth of ½ inch (1cm), or June–August in a cold frame or cool greenhouse at the same depth. Germination takes 21–35 days. When

the seedlings are large enough to handle, line them out into deep boxes and grow them on until the following spring. Transfer the plants to their permanent positions during April, spacing them 24 inches (61cm) apart.

The flowers bloom June–July.

Kochia scoparia 'Trichophylla'
SUMMER CYPRESS, BURNING BUSH

Hardy annual
Native to Europe

A rapid-growing foliage plant which forms a bush of about 12 inches (30cm) in diameter. The foliage is pale green during the summer and changes to a coppery-crimson during the autumn. The plant grows to a height of 24–36 inches (61–91cm). It grows best in a sunny position and on a light soil.

Sow the seeds March–April at a temperature of 60–68°F (15–20°C) and at a depth of ⅛ inch (3mm) in boxes of seed compost. Germination takes 5–7 days. When the seedlings are large enough to handle, prick them out into boxes and plant them out in the garden

Above: Kochia scoparia 'Trichophylla' forms an excellent backcloth to a border planting.

Left: Kniphofia: its brilliant colours are aptly matched by one of its common names, Red Hot Poker.

during May, spacing them 12 inches (30cm) apart. Alternatively, sow the seeds April–May in the open ground at a depth of $\frac{1}{4}$ inch (6mm). When the seedlings are large enough to handle, thin them out to 12 inches (30cm) apart.

Lagerstroemia indica

Half-hardy shrub
Native to China

This deciduous shrub bears open single flowers which vary in colour from pink to deep red. It may grow to a height of 5 feet (152cm). Although it is usually grown indoors as a pot plant, it may be planted outside in a sheltered garden.

Sow the seeds January–April at a temperature of 68–78°F (20–25°C) and at a depth of $\frac{1}{8}$ inch (3mm). Germination is sporadic. When the seedlings are large enough to handle, prick them out singly into 3$\frac{1}{2}$-inch (9-cm) pots of potting compost. When the plants reach a height of 3–4 inches (8–10cm), pinch out the leading shoots. Pot on as necessary.

Plants grown outside may require protection during the winter months.

The flowers bloom August–October.

Larkspur see *Consolida ajacis*

Lathyrus

L. latifolius
EVERLASTING SWEET PEA

Hardy perennial
Native to Europe

This climber is a favourite in cottage gardens. It has small rose, red, and white flowers and grows to a height of 6–10 feet (1.8–3 metres). It is widely grown as a covering plant on walls and fences. It thrives in any fertile well-drained garden soil and requires a sunny position.

Sow the seeds March–April at a depth of $\frac{1}{2}$ inch (1cm) in the flowering site. Germination takes 14–28 days. When the seedlings are large enough to handle, thin them out to 18 inches (46cm) apart.

The flowers bloom June–September.

L. odoratus
SWEET PEA

Hardy annual
Native to Sicily

This climbing plant has delicately scented flowers in many colours including shades of red, blue, pink, and purple as well as white. It grows to a height of 6–10 feet (1.8–3 metres). Best results are obtained if it is grown on a well-drained loam containing bulky organic matter. It requires a sunny position.

Sow the seeds January–March at a temperature of 60–68°F (15–20°C) and at a depth of $\frac{1}{2}$ inch (1cm). Alternatively, sow the seeds April–May in open ground at a depth of $\frac{1}{2}$ inch (1cm). In order to have early blooms the following year, sow the seeds September–October and over-winter them in a cold frame or greenhouse. Do not water the seeds for 2 days after sowing, but then thoroughly soak them and keep the sowing medium moist until germination has taken place. Germination takes 7–21 days. When the seedlings are planted in the flowering position, space them 10 inches (25cm) apart.

The flowers bloom June–September.

Lavandula spica
OLD ENGLISH LAVENDER

Hardy evergreen shrub
Native to the Mediterranean region

This shrub has silver-grey leaves with spikes of blue-grey flowers which

Sweet peas (*Lathyrus odorata*) are a favourite
summer flower. There are numerous varieties
available.
Above: 'Early Multiflora Gigantea Mixed'.
Above left: 'Elisabeth Collins'.
Left centre: 'Noel Sutton'.
Left below: 'Little Elfin Mixed'.
Below: 'Cream Beauty'.

bloom during the summer. It grows to a height of 36–48 inches (91–122cm). It succeeds best on a light sandy soil and requires an open sunny position. It may be used as a hedging plant.

Sow the seeds April–May in shallow drills at a depth of ¼ inch (6mm) in a seed bed. Germination is sporadic, taking 28–42 days. When the seedlings are large enough to handle, thin them out to 12 inches (30cm) apart. Transfer the plants to their permanent positions during the autumn, spacing them 18–24 inches (46–61cm) apart.

Lavender flowers may be dried and used in sachets and pot-pourri.

Lavatera trimestris
Lavatera rosea
MALLOW

Hardy annual
Native to south-west Europe

A bushy plant with pale-green foliage and rose-pink flowers. It grows to a height of 24–36 inches (61–91cm). Mallow is a popular border plant and

may be used also to make a temporary hedge. Most soils are suitable for Mallow and it requires a sunny position.

Sow the seeds March–May at a depth of ½ inch (1cm) where the plants are to flower. Germination takes 21–35 days. When the seedlings are large enough to handle, thin them out to 24 inches (61cm) apart.

The flowers bloom July–September.

Lavender, Sea see *Limonium latifolium* and *Limonium sinuatum*

Right: Lavatera trimestris 'Silver Cup'.
Below right: L. trimestris 'Mont Blanc'.
Below: L. trimestris 'Suttons Loveliness'.

A cottage garden planted with a medley of colourful plants including pinks (*Dianthus caryophyllus*), pot marigolds (*Calendula officinalis*) and hollyhocks (*Althaea rosea*).

Leontopodium alpinum

EDELWEISS

Hardy perennial
Native to the European Alps

This plant has grey-green foliage, and
flowerheads which are covered with
white cottony down. It grows to a
height of 8 inches (20cm). Edelweiss is
especially suitable for the rockery. It
does best on a well-drained sandy soil
and requires a sunny position.

Sow the seeds March–April in a cold
frame or greenhouse. Make up the
compost mixture of 2 parts potting
compost No 1 to 1 part grit, and sow
the seeds at a depth of $\frac{1}{4}$ inch (6mm).
Germination takes 14–28 days. When
the seedlings are large enough to
handle, prick them out into boxes.
When the plants are well developed,
transfer them to 3-inch (8-cm) pots
and grow them on. Transfer the plants
to their permanent positions during
the following spring, spacing them far
enough apart to allow for their spread
of up to 9 inches (23cm).

The flowers bloom July–September.

Leptosiphon hybridus see Gilia
lutea

Levisticum officinale

LOVAGE

Hardy perennial
Native to the Mediterranean region

The leaves of this plant resemble those
of Parsley and it grows to a height of
48 inches (122cm). It requires a rich
soil and a sunny position.

Sow the seeds April–May in shallow
drills in a seed bed. When the
seedlings are large enough to handle,
thin them out to 12 inches (30cm)
apart. Transfer the plants to their
permanent positions during the
autumn, spacing them 36 inches
(91cm) apart.

The leaves may be chopped and
used in broths, stews, or salads.

Liatris pycnostachya

BLAZING STAR, GAYFEATHER

Hardy perennial
Native to North America

This species has mid-green foliage and
spikes of purple thistlelike flowerheads.
It grows to a height of 24–36 inches
(61–91cm). It is an excellent border
plant as well as being popular as a cut
flower. Most soils are suitable for this
plant but it requires a sunny position.

Sow the seeds April–September at a
depth of $\frac{1}{4}$ inch (6mm) in a cold frame
or greenhouse. Germination takes
14–21 days. When the seedlings are
large enough to handle, prick them out
into boxes. When the plants are well
developed, transfer them to a nursery
bed, spacing them about 12 inches
(30cm) apart. Transfer the plants to
their permanent positions during the
autumn of the following year, spacing
them 12–18 inches (30–46cm) apart.

The flowers bloom July–September.

Lilium regale

THE REGAL LILY

Hardy bulbous perennial
Native to China

This Lily has trumpet-shaped blooms
which are flushed inside with yellow

Lilium regale.

shading to white towards the outer
edge; outside they are streaked with
brown. It grows to a height of 42
inches (107cm). Most soils are suitable
for this plant, but it needs a sunny
position.

Sow the seeds thinly at a depth of $\frac{1}{2}$
inch (1cm) in pots during the autumn
and leave them out of doors in a cold
frame during the winter. Germination
takes 21–42 days. Bring them in to
warmer conditions during the spring.
Keep the seedlings in the pots for two
seasons. At the end of the second
season they may either be lined out in
nursery beds or placed in their
permanent positions. In either case,
space the plants 12 inches (30cm)
apart. Many of them will flower during
their third year.

The flowers bloom during July.

Lily, Peruvian see *Alstroemeria ligtu
hybrids*

Lily, Torch see *Kniphofia*

Limnanthes douglasii
POACHED EGG PLANT

Hardy annual
Native to the western USA

This easily grown plant has bright-
green foliage and saucer-shaped
flowers which are yellow edged with
white. It grows to a height of 6 inches
(15cm) and requires an open sunny
position and a well-drained soil.

Sow the seeds March–May at a
depth of $\frac{1}{8}$ inch (3mm) in the flowering
site. Germination takes 14–21 days.
When the seedlings are large enough
to handle, thin them out to 4 inches
(10cm) apart. Seeds sown during
September will flower the following
year, but may need to be protected by
cloches in cold areas.

The flowers bloom June–August.

Limnanthes douglasii is very easy to grow,
bringing a splash of yellow and white to the
summer bedding scheme.

Limonium

L. latifolium
Statice latifolia
SEA LAVENDER

Hardy perennial
Native to southern Europe

This species has large heads of
lavender-blue flowers and deep-green
foliage. Like *L. sinuatum*, it may be
dried for use in winter floral
arrangements. It grows to a height of
24 inches (61cm). It requires a sunny
position and well-drained soil.

Sow the seeds April–May at a depth
of $\frac{1}{4}$ inch (6mm) in a cold frame or
greenhouse. Germination takes 14–28
days. When the seedlings are large
enough to handle, prick them out into
boxes and grow them on. Plant them
out in their permanent positions
September–October, spacing them 18
inches (46cm) apart.

The flowers bloom July–September.

L. sinuatum
Statice sinuata
SEA LAVENDER

Hardy perennial usually grown as a
half-hardy annual
Native to the Mediterranean region

The plant has attractive sprays of
'everlasting' flowers in yellow, pink,
lavender, dark blue, and white. It is

Limonium sinuatum 'Sinuata Special Mixture'
(*above*) and 'Suworowii' (*below*).

popular with flower arrangers as it
may be dried and used in winter floral
decorations. The plant grows to a
height of 18 inches (46cm). It succeeds
on most garden soils, but requires an
open sunny position.

Sow the seeds January–March at a
temperature of 60–68°F (15–20°C)
and at a depth of $\frac{1}{4}$ inch (6mm).
Germination takes 10–21 days. When
the seedlings are large enough to
handle, prick them out into boxes and
harden them off. Transfer them to
their flowering positions during May,
spacing them 12 inches (30cm) apart.

The flowers bloom July–September.

Linaria maroccana
TOADFLAX

Hardy annual
Native to North Africa

This compact bushy plant has dainty
spikes of snapdragonlike flowers in
various colours including violet, blue,
yellow, red, and pink. It grows to a
height of 8–15 inches (20–38cm).
Linaria is a good edging and border
plant. It requires a sunny position and

Linaria maroccana 'Fairy Bouquet'. This
species resembles a miniature Snapdragon and
its wide range of bright colours and compact
habit make it excellent for use as an edging
plant or to bring a bright splash of colour to the
border.

grows on most types of soil.

Sow the seeds March–May at a depth of ⅛ inch (3mm) in the flowering site. Germination takes 10–14 days. When the seedlings are large enough to handle, thin them out to 6 inches (15cm) apart.

The flowers bloom June–July.

Linum
FLAX

L. grandiflorum
SCARLET FLAX

Hardy annual
Native to North Africa

This species of Flax has pale-green foliage and rose-coloured saucer-shaped flowers. The variety 'Rubrum' has bright-crimson flowers. The plant grows to a height of 12 inches (30cm). Any ordinary garden soil is suitable provided that it is well-drained. This plant requires a sunny position.

Sow the seeds March–May at a depth of ¼ inch (6mm) in the flowering site. Germination takes 14–28 days. When the seedlings are large enough

Linum grandiflorum 'Grandiflorum rubrum'.

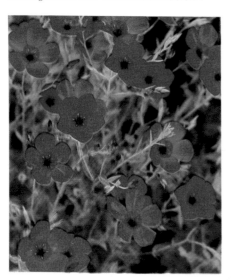

to handle, thin them out to 5 inches (13cm) apart.

The flowers bloom June–August.

L. narbonense
Hardy perennial
Native to southern Europe

This species has grey-green leaves and sky-blue flowers. It grows to a height of 24 inches (61cm). Like *L. grandiflorum*, it requires a well-drained soil and sunny position.

Sow the seeds June–July at a depth of ¼ inch (6mm) in a seed bed. Germination takes 14–28 days. When the seedlings are large enough to handle, thin them out to 9 inches (23cm) apart and grow them on. Transfer the plants to their permanent positions during October, spacing them 12 inches (30cm) apart.

The flowers bloom June–September.

Lithops
LIVING STONES

Greenhouse perennial
Native to South Africa

A group of lithops showing some of the varieties available.

The seeds are usually supplied as a pack of mixed varieties by seedsmen. Lithops is a very small succulent plant. It consists of two tightly fused, fleshy leaves which are conical or round in shape and resemble pebbles. The stemless daisylike flowers emerge from the cleavage of the leaves. The plant grows to a height of about 1 inch (2.5cm).

Sow the seeds at any time at a temperature of 50–60°F (10–15°C). Just press them into the surface of a sandy compost in 3-inch (8-cm) pots. Do not water the plants between October and April as they are resting during that period.

Two compact varieties of *Lobelia erinus*: 'Crystal Palace' (*above*) and 'String of Pearls' (*below*).

Lobelia erinus

Half-hardy perennial usually grown as a half-hardy annual
Native to South Africa

The foliage is light green and the flowers are blue, white, or red. Trailing or Pendula varieties are also available and are particularly suitable for hanging-baskets. Lobelias grow to a height of 4–9 inches (10–23cm). These dwarf spreading plants are usually grown for edgings.

Sow the seeds January–March at a temperature of 60–68°F (15–20°C). As the seeds are very small, sow them thinly and just press them into the surface of the compost. It is unnecessary to cover them, but keep the compost moist. Germination takes 7–14 days. When the seedlings are large enough to handle, prick them out into boxes and harden them off. Transfer the plants to the flowering site during May, spacing them 4 inches (10cm) apart.

The flowers bloom from May until the first autumn frosts.

Lobularia maritima see *Alyssum maritimum*

Lovage see *Levisticum officinale*

Love-in-a-mist see *Nigella damascena*

Love-lies-bleeding see *Amaranthus caudatus*

Lunaria annua
HONESTY

Hardy biennial
Native to Europe

The flowers are purple and the foliage is mid-green. Honesty grows to a

height of 24 inches (61cm). It grows
best on a light soil and in a shaded
position. This plant is grown for its
beautiful seed pods which are like flat
silver discs and are much used in
winter floral arrangements.

Sow the seeds May–June in a
nursery bed at a depth of $\frac{1}{2}$ inch (1cm).
Germination takes 14–21 days. When
the seedlings are large enough to
handle, thin them out to 6 inches
(15cm) apart and grow them on until
the autumn. Transfer the plants to
their flowering positions during
September, spacing them 12 inches
(30cm) apart.

The flowers bloom April–June.

Lupinus
LUPIN

Hardy perennial
Derived from species native to the western USA

Most Lupins grown today are hybrids
known as 'Russell' Lupins. The upper
petals of Lupin flowers are turned
back and the lower petals are
compressed. The flowers are carried
on spikes which grow to a height of 24
inches (61cm). The foliage is mid-

Lupins are handsome plants, bringing height to
the border. They dislike limy soil.

green. The plant grows to a height of
36 inches (91cm). It grows in either a
sunny or shady position and requires
an acid or neutral soil. The plant lasts
longer if it is grown on a light soil.

Sow the seeds April–June at a depth
of $\frac{1}{2}$ inch (1cm) in a cold frame or
greenhouse. Germination takes 21–28
days. Transfer the plants June–August
to their permanent positions, spacing
them 24 inches (61cm) apart. They
will begin to flower the following year.

The flowers bloom May–July.

Lychnis
CAMPION

L. arkwrightii

Hardy perennial

This hybrid form of Lychnis has mid-
green leaves and flowers which range
from orange-red to salmon-pink in
colour. It grows to a height of 18
inches (46cm). Any light well-drained
garden soil is suitable for this plant
and it grows in either a sunny or
partially shaded position.

Sow the seeds March–May at a
depth of $\frac{1}{4}$ inch (6mm) in the flowering

A group of lupins from the 'Russell' Strain
showing the wide range of colours available.

site. Germination takes 14–28 days. When the seedlings are large enough to handle, thin them out to 9 inches (23cm) apart. Seeds sown early in the year will flower during their first year.

The flowers bloom June–August.

L. chalcedonica
JERUSALEM CROSS

Hardy perennial
Native to Russia

This species has dense heads of scarlet cross-shaped flowers. It grows to a height of 36 inches (91cm). It requires the same conditions as *L. arkwrightii* and the sowing instructions are the same. When the plants are thinned out, however, space them 12–15 inches (30–38cm) apart.

The flowers bloom June–July.

L. coronaria
Agrostemma coronaria

Hardy perennial
Native to eastern Europe

The foliage is silver in colour and has a woolly texture; the flowers are rich crimson. The plant grows to a height of 24 inches (61cm). Any well-drained soil is suitable for this plant and it grows either in full sun or partial shade.

Sow the seeds during February at a temperature of 55–61°F (13–16°C). Germination takes 10–14 days. When the seedlings are large enough to handle, prick them out into boxes of potting compost and harden them off. Transfer the plants to their permanent positions during May, spacing them 9–12 inches (23–30cm) apart. Alternatively, sow the seeds May–July out of doors for flowering the following summer. When the seedlings are large enough to handle, thin them out to 9–12 inches (23–30cm) apart.

The flowers bloom July–September.

L. viscaria
Viscaria vulgaris

Hardy annual
Native to Europe

The foliage is bright green; the single five-petalled flowers have dark eyes and are usually pastel coloured in shades of white, pink, and mauve. The plant grows to a height of 12 inches (30cm). It requires a well-cultivated soil and a sunny position.

Sow the seeds March–May in the flowering site at a depth of $\frac{1}{4}$ inch (6mm). Germination takes 7–10 days. When the seedlings are large enough to handle, thin them out to 2–3 inches (5–8cm) apart.

The flowers bloom May–July.

Lychnis viscaria 'Suttons Brilliant Mixture': a very free-flowering variety which looks best when planted in patches or strips in the border.

Malcolmia maritima
VIRGINIAN STOCK

Hardy annual
Native to the Mediterranean region

Virginian Stock has grey-green leaves and white cross-shaped flowers. It grows to a height of about 12 inches

(30cm). Garden varieties are available in various colours including red, yellow, white, and lilac. Dwarf forms reach a height of 6 inches (15cm). This plant may be grown as an edging plant or in a rock garden. It grows almost anywhere but the best results are obtained if it is given a sunny position.

Sow the seeds March–May at a depth of $\frac{1}{4}$ inch (6mm) in the flowering site. Germination takes 7–10 days. When the seedlings are large enough to handle, thin them out to about 6 inches (15cm) apart.

The flowers bloom about 4 weeks after sowing and will continue to bloom for about 6–8 weeks.

Mallow see *Lavatera trimestris*

Marigold see *Tagetes tenuifolia pumila*

Marigold, African see *Tagetes erecta*

Marigold, English see *Calendula officinalis*

Marigold, French see *Tagetes patula*

Marigold, Pot see *Calendula officinalis*

Marjoram see *Origanum*
Marjoram, Knotted see *Origanum majorana*

Marjoram, Pot see *Origanum onites*

Marjoram, Sweet see *Origanum majorana*

Marrubium vulgare
WHITE HOREHOUND

Hardy perennial
Native to Europe

The thyme-scented leaves are covered with white hairs. The flowers, which bloom July–September, are also white and are attractive to bees. It requires full sun and grows easily on any poor, dry soil. It is especially suitable for chalky soils.

Sow the seeds during April in the open ground at a depth of $\frac{1}{2}$ inch (1cm). When the seedlings are about 4 inches (10cm) tall thin them out to 10 inches (25cm) apart.

White Horehound may be used to make a syrup to relieve a persistent cough. It has been used for many centuries to make candy to relieve throat and bronchial congestion.

Marvel of Peru see *Mirabilis jalapa*

Mask Flower see *Alonsoa warscewiczii*

Matricaria exima see *Chrysanthemum parthenium*

Matthiola

M. bicornis
NIGHT-SCENTED STOCK

Hardy annual
Native to south-east Europe

The long narrow leaves are greyish-green. The purple-lilac flowers, which open at night, have four petals. The plant grows to a height of 12–15 inches (30–38cm). It grows well on any good garden soil and tolerates partial shade. As the flowers open during the evening, this plant is often grown under windows so that its perfume will waft into the house.

Sow the seeds March–May at a depth of $\frac{1}{4}$ inch (6mm) in the flowering site. Germination takes 10–14 days. When the seedlings are large enough to handle, thin them out to 9 inches (23cm) apart.

The flowers bloom July–August.

M. incana
STOCK
Native to Europe

Stocks have spikes of clove-scented flowers which are either richly coloured or in pastel shades. The garden varieties are divided into the following groups:

Ten Week Stock 'Giant Perfection Mixed'

Brompton Stock 'Spring-Flowering'.

Ten Week Stock

This group of hardy annuals contains the following strains: Dwarf varieties, which grow to a height of 12 inches (30cm); Excelsior varieties, which grow to a height of 30 inches (76cm); and Mammoth or Beauty varieties, which grow to a height of 18 inches (46cm). All the plants in this group require a sunny position and well-cultivated soil containing a little lime.

Sow the seeds during April at a depth of $\frac{1}{4}$ inch (6mm) in the flowering position. Germination takes 7–10 days. When the seedlings are large enough to handle, thin them out to 12 inches (30cm) apart.

The flowers bloom July–September.

Perpetual Flowering Stock

This dwarf plant is a hardy annual and grows to a height of 15 inches (38cm). It has large spikes of white flowers. Like all Stocks, it requires the same conditions as those for Ten Week Stock.

The sowing instructions are the same as those for Ten Week Stock.

The flowers bloom June–July.

Brompton Stock

This is a hardy biennial with a bushy habit. It bears both single and double flowers and grows to a height of 18 inches (46cm). The colours available include red, pink, yellow, purple, and white. It requires the same conditions as those for Ten Week Stock.

Sow the seeds June–July at a depth of $\frac{1}{4}$ inch (6mm) in a well-prepared seed bed. Germination takes 7–10 days. When the seedlings are large enough to handle, thin them out to 12 inches (30cm) apart. Transfer the plants during October to the flowering site, spacing them 12 inches (30cm) apart.

The flowers bloom June–July.

East Lothian Stock

This plant may be grown as an annual or biennial. It grows to a height of about 15 inches (38cm). The colour range is the same as that for Brompton Stock. This strain requires the same conditions as those for Ten Week Stock.

Sow the seeds February–March at a temperature of 60–68°F (15–20°C) and at a depth of $\frac{1}{4}$ inch (6mm). Germination takes 7–10 days. When the seedlings are large enough to handle, prick them out into boxes and harden them off. Transfer the plants to their flowering positions during May, spacing them 12 inches (30cm) apart.

The flowers bloom from July until the first autumn frosts.

Trysomic Stock

This annual Stock produces a high proportion (85 per cent) of double flowers. The plant grows to a height of 12–20 inches (30–51cm), depending on the variety. It requires the same conditions as those for Ten Week Stock.

'Giant Excelsior Mixed' another Ten Week variety (*above*) ; 'Giant Imperial Mixed' a Trysomic Stock (*below*).

Sow the seeds February–March at a temperature of 55–60°F (13–15°C) and at a depth of ¼ inch (6mm). Germination takes 7–10 days. When

the seedlings are large enough to handle, prick them out into boxes. At this point, discard the weaker seedlings as they will become single-flowered plants. The more vigorous seedlings will become double-flowered plants. Grow the plants on and harden them off. Transfer them to the flowering site during May, spacing them 12 inches (30cm) apart.

The flowers bloom June–July.

Meconopsis

M. betonicifolia
M. Baileyi
HIMALAYAN BLUE POPPY

Hardy perennial
Native to Tibet

This species has brilliant Cambridge-blue flowers. It grows to a height of 36 inches (91cm). It requires a rich loamy soil and grows best in a shady position.

Sow the seeds during December, at a depth of ⅛ inch (3mm) in pots of seed compost. Leave them out of doors until February and then bring them inside to warmer conditions. Germination seems to be helped by exposing the seeds to frost, but is sporadic, taking from 28 days onwards. When the seedlings are large enough to handle, prick them out singly into 3-inch (8-cm) pots and grow them on until the autumn. Transfer the plants to their permanent positions September–October, spacing them 12–18 inches (30–46cm) apart.

The flowers bloom June–July.

M. cambrica
WELSH POPPY

Hardy perennial
Native to Europe

This plant has bright-yellow poppylike flowers and grows to a height of 18 inches (46cm). It thrives

The Welsh poppy, *Meconopsis cambrica* is suitable for the rock garden.

on any type of soil and is suitable for any position in the garden.

Sow the seeds May–June at a depth of $\frac{1}{4}$ inch (6mm) in a seed bed. When the seedlings are large enough to handle, thin them out to 6 inches (15cm) apart and grow them on until the autumn. Transfer the plants to their permanent positions September–October, spacing them 12 inches (30cm) apart.

The flowers bloom May–June.

Melissa officinalis
BALM

Hardy perennial
Native to central and southern Europe

A bushy plant with lemon-scented heart-shaped leaves. It grows to a height of 24–48 inches (61–122cm). Balm grows best on a light soil and needs a sunny position.

Sow the seeds April–May in shallow drills at a depth of $\frac{1}{4}$ inch (6mm) in a seed bed. Germination takes 14–21 days. When the seedlings are large enough to handle, transplant them to their permanent positions, spacing them 18–24 inches (46–61cm) apart.

The leaves may be used fresh or dried as a herb tea, in fish and chicken dishes, and in salad dressings.

Mentzelia lindleyi resembles St John's wort (*Hypericum calycinum*).

Mentzelia lindleyi

Bartonia aurea

Hardy annual
Native to western North America

This plant has succulent stems and bright yellow flowers. It grows to a height of 18 inches (46cm). It thrives on any soil, but a sunny position is preferred.

Sow the seeds March–May at a depth of $\frac{1}{4}$ inch (6mm) in the flowering site. Germination takes 7–10 days. When the seedlings are large enough to handle, thin them out to 9 inches (23cm) apart.

The flowers bloom June–August.

Mesembryanthemum

M. criniflorum
LIVINGSTONE DAISY

Half-hardy perennial usually grown as a half-hardy annual
Native to South Africa

This plant has a dwarf spreading habit. The succulent leaves are light green and the daisylike flowers, which are produced in profusion, are white edged with rose, apricot, crimson, or pink. It grows to a height of 4–6 inches

Mesembryanthemum criniflorum 'Suttons Sparkles'.

(10–15cm). It is suitable for rockeries and for use as an edging plant. Although it grows on most soils, it requires a sunny position.

Sow the seeds February–April at a temperature of 60–68°F (15–20°C) and at a depth of $\frac{1}{8}$ inch (3mm). Germination takes 14–21 days. When the seedlings are large enough to handle, prick them out into boxes and harden them off. Transfer the plants to their permanent positions May–June, spacing them 12 inches (30cm) apart.

The flowers bloom June–August.

M. tricolor

Half-hardy annual
Native to South Africa

Like M. criniflorum, this plant has a dwarf spreading habit. The succulent leaves are dark green and the flowers are deep rose and white with dark centres. It grows to a height of 3 inches (8cm). It thrives on most soils and requires a sunny position.

The sowing instructions are the same as those for M. criniflorum.

The flowers bloom June–August.

Mignonette see Reseda odorata

Milkweed see Asclepias curassavica

Mimosa see Acacia dealbata

Mimosa-leaved Ebony Tree see Jacaranda ovalifolia

Mimosa pudica
SENSITIVE PLANT

Greenhouse shrub
Native to the West Indies

This plant has fernlike leaves which, when touched, fold up. It has pink ball-like flowers, but is grown primarily for its decorative foliage. Mimosa grows to a height of 24 inches (61cm). It is usually grown as a pot plant in the greenhouse or conservatory.

Sow the seeds at any time at a temperature of 68–78°F (20–25°C) and at a depth of $\frac{1}{4}$ inch (6mm). Germination takes 14–28 days. When the seedlings are large enough to handle, prick them out singly into 3-inch (8-cm) pots of potting compost. Pot on as necessary until the plants are in 5-inch (13-cm) pots.

The flowers bloom July–August.

Mimulus
MONKEY FLOWER

Half-hardy annual
Native to the Americas

The trumpet-shaped flowers are gaily marked and have spotted throats. The plant grows to a height of 9 inches (23cm). As Mimulus grows in boggy conditions in the wild, ideally it should be planted in a moist site in the garden.

Sow the seeds February–May at a temperature of 60–68°F (15–20°C). Just press the seeds into the surface of the seed compost. Germination takes 10–14 days. When the seedlings are large enough to handle, prick them out into boxes and harden them off.

Two varieties of *Mimulus*: 'Suttons Giant Mixed' (*above*) and 'Royal Velvet', an F₁ hybrid (*below*).

yellow, red, or white. The flowers usually open during the afternoon. The plant grows to a height of 24 inches (61cm). It is usually grown as a large pot plant in the greenhouse, but it may be grown out of doors in a warm sheltered position. If it is to be grown outside, it prefers a light, fairly rich soil.

Sow the seeds February–March at a temperature of 60–68°F (15–20°C) and at a depth of ¼ inch (6mm). Germination takes 14–21 days. When the seedlings are large enough to handle, prick them out into boxes. Plant those that are to be grown as pot

Mirabilis jalapa is often called the Four o'Clock Plant as its flowers open during the afternoon.

Transfer the plants to their flowering positions during May, spacing them 9–12 inches (23–30cm) apart. The flowers bloom June–September.

plants in 6-inch (15-cm) pots of potting compost; harden off those that are to be planted outside and transfer them to their flowering positions at the end of May, spacing them 12 inches (30cm) apart.

The flowers bloom June–September.

Mirabilis jalapa
MARVEL OF PERU

Perennial grown as a half-hardy annual
Native to tropical America

This plant has mid-green leaves and trumpet-shaped flowers in shades of

Molucella laevis
BELLS OF IRELAND

Hardy annual
Native to the Levant

This species has graceful stems covered with green shell-like sheaths which are delicately veined. The flowers are white. It grows to a height of 36 inches (91cm). This plant is a great favourite with flower arrangers as it dries well and can therefore be used in both summer and winter floral decorations. Ideally, it requires a sunny position and light soil.

Sow the seeds February–March at a temperature of 60–68°F (15–20°C) and at a depth of $\frac{1}{4}$ inch (6mm). Germination is sporadic, taking from 21 days onwards. When the seedlings are large enough to handle, prick them out into boxes and harden them off in a cold frame. Transfer the plants to their flowering positions during May, spacing them 9 inches (23cm) apart.

The flowers bloom during the summer.

Monarda didyma
BERGAMOT, OSWEGO TEA, BEE BALM

Hardy perennial
Native to the eastern USA and Canada

The mid-green leaves are hairy. The bright scarlet flowers, which bloom June–September, are very attractive to bees and butterflies. Garden varieties are available in many colours including rose-pink, white, and violet. It grows to a height of 24–36 inches (61–91cm). It requires a moist soil and grows in either a sunny or partially shaded position.

Sow the seeds during March at a temperature of 60°F (15°C). Use seed compost and sow the seeds at a depth of $\frac{1}{4}$ inch (6mm). When the seedlings are large enough to handle, prick them out and grow them on in nursery rows outside. Space them 9 inches (23cm) apart in the rows. Transplant them to their permanent positions during October, spacing them 15 inches (38cm) apart.

Bergamot leaves are used to make Oswego tea, and both leaves and flowers may be eaten in salads.

Monkey Flower see *Mimulus*

Morning Glory see *Pharbitis tricolor*

Mullein see *Verbascum phoeniceum*

Myosotis sylvatica
FORGET-ME-NOT

Hardy biennial
Native to Europe

The fragrant misty-blue flowers blend well with all the other spring flowers. The hairy leaves are mid-green. The plant grows to a height of 12 inches (30cm). It thrives in almost any position in the garden but is particularly suitable for areas of partial shade. Moist conditions suit it best.

Sow the seeds May–July in a seed bed at a depth of $\frac{1}{4}$ inch (6mm). Germination takes 14–21 days. When the seedlings are large enough to handle, thin them out to 6 inches (15cm) apart. Transfer the plants to their flowering site during September, spacing them 6 inches (15cm) apart.

The flowers bloom May–June.

Myrrhis odorata
SWEET CICELY

Hardy perennial
Native to northern Europe

The leaves of this plant are bright green and resemble those of the Oak Fern. It is slow growing but eventually reaches a height of 5 feet (1.5 metres). It requires a rich soil which contains plenty of humus.

Sow the seeds April–May in shallow drills where the plants are to be permanently sited. When the seedlings are large enough to handle, space them 24 inches (61cm) apart.

The leaves, which have a slight aniseed flavour, may be used in salads.

Nasturtium see *Tropaeolum majus*

Nemesia strumosa 'Carnival Mixture' (*above*) and 'Sparklers' (*below*). Nemesias are easy to grow and quick-flowering.

Nemesia strumosa

Half-hardy annual
Native to South Africa

This compact plant produces many flowerheads of brilliant colours including yellow, blue, scarlet, rose-pink, orange, and cherry-red. The plant grows to a height of 8–18 inches (20–46cm), depending on the variety. It may be used as a summer bedding plant or for edging. Plant Nemesias on soil which contains plenty of humus and in an open sunny position.

Sow the seeds February–April at a temperature of 60–68°F (15–20°C) and at a depth of ⅛ inch (3mm). When the seedlings are large enough to handle, prick them out into boxes and harden them off. Transplant them to their flowering positions May–June, spacing them 4–6 inches (10–15cm) apart. Nemesias may be sown at a depth of ⅛ inch (3mm) July–October for winter flowering in a greenhouse.

The flowers bloom June–August.

Nemophila menziesii
Nemophila insignis
BABY BLUE EYES

Hardy annual
Native to the western USA

The leaves are feathery and light green; the buttercup-shaped flowers are sky-blue. The plant grows to a height of 9 inches (23cm). It thrives in cool moist situations and may be grown in either a sunny or partially shaded position.

Sow the seeds March–May at a depth of ¼ inch (6mm) in the flowering

Nemophila menziesii can be planted in cool moist situations which are unsuitable for many other plants.

site. Germination takes 10–14 days. When the seedlings are large enough to handle, thin them out to 6 inches (15cm) apart.

The flowers bloom June–August.

Nepeta × faassenii
Nepeta mussini
CATMINT

Hardy perennial
Native to the Caucasus

This plant has blue-grey foliage and spikes of mauve flowers. It grows to a height of 9 inches (23cm). It may be used for edging or as a border plant. It is best suited by a sunny or partially shaded position and well-drained soil.

Sow the seeds May–June at a depth of $\frac{1}{4}$ inch (6mm) in a nursery bed. Germination takes 14–28 days. When the seedlings are large enough to handle, thin them out to 6 inches (15cm) apart and grow them on until the autumn. Transfer the plants to their permanent positions September–October, spacing them 12 inches (30cm) apart. The first flowers will appear the following year.

The flowers bloom May–September.

Nicandra physaloides
APPLE OF PERU, SHOO FLY PLANT

Hardy annual
Native to Peru

This plant has pale-blue bell-shaped flowers which are followed by small apple-shaped fruits. It grows to a height of 36 inches (91cm). It is often planted beneath windows as it is said to repel flies. It requires a sunny position and rich moist soil.

Sow the seeds February–March at a temperature of 50–60°F (10–15°C) and at a depth of $\frac{1}{8}$ inch (3mm). Germination will take 14–21 days. When the seedlings are large enough

to handle, prick them out into boxes and harden them off in a cold frame. Transfer them to their flowering positions during May, spacing them 12 inches (30cm) apart. Alternatively, sow the seeds April–May in the flowering site at a depth of $\frac{1}{4}$ inch (6mm). When the seedlings are large enough to handle, thin them out to 12 inches (30cm) apart.

The flowers bloom July–September.

Nicotiana
TOBACCO PLANT
Native to Brazil

N. alata
N. affinis

Half-hardy annual

This variety has mid-green foliage and white tube-shaped flowers. The flowers may grow to a length of 3 inches (8cm). The plant is sticky to touch and grows to a height of 24–36 . inches (61–91cm). Cultivated varieties

Nicotiana alata 'Nicki': an F₁ hybrid.

Above: another F₁ hybrid 'Crimson Rock'; (*below*) the green-flowered variety 'Lime Green'.

the seedlings are large enough to handle, prick them out into boxes and harden them off. Plant them out in the flowering site from late May onwards, spacing them about 12 inches (30cm) apart.

The flowers bloom June–September.

N. sylvestris

Half-hardy biennial

This plant has very large, pale-green leaves and drooping clusters of tube-shaped white flowers. The flowers close in full sunshine but remain open on dull days. The plant grows to a height of 36–48 inches (91–122cm).

The sowing instructions are the same as those for *N. affinis*, but when the plants are transferred to their flowering positions, space them 24 inches (61cm) apart.

The flowers bloom during June.

Nigella damascena
LOVE-IN-A-MIST

Hardy annual
Native to Europe

Nigella damascena 'Miss Jekyll'.

are available in many colours including cream, yellow, pink, crimson, and white. The variety 'Lime Green' has yellow-green flowers. Tobacco Plants prefer an open position, but are tolerant of partial shade. They require a rich well-cultivated soil.

Sow the seeds February–April at a temperature of 60–68°F (15–20°C) and at a depth of ⅛ inch (3mm). Germination takes 10–14 days. When

Nigella damascena 'Persian Jewel'.

This plant has bright green fernlike foliage and blue or white flowers which resemble Cornflowers. Cultivated varieties are available with mauve, purple, and rose-pink flowers as well as blue and white ones. The plant grows to a height of 18–24 inches (46–61cm). The brown seed-pods are much appreciated by flower arrangers for winter decorations. Nigella requires a sunny position and well-drained soil.

Sow the seeds March–May at a depth of $\frac{1}{4}$ inch (6mm) in the flowering site. Germination takes 14–21 days. When the seedlings are large enough to handle, thin them out to 6–9 inches (15–23cm) apart. Seeds sown during September will withstand the winter and flower early the following summer.

The flowers bloom June–August.

Ocimum
BASIL
Native to tropical Asia

O. basilicum
SWEET BASIL

Half-hardy annual

This herb has shiny green aromatic leaves. It grows to a height of 12–18 inches (30–46cm). It requires a position in full sun and light well-drained soil.

The plant does not take kindly to having its roots disturbed and the best results are obtained by sowing the seeds during May at a depth of $\frac{1}{4}$ inch (6mm) in the flowering site. Germination takes 14–21 days. When the seedlings are large enough to handle, thin them out to 15 inches (38cm) apart. Alternatively, sow the seeds during March at a temperature of 50–60°F (10–15°C) and at a depth of $\frac{1}{4}$ inch (6mm). When the seedlings are large enough to handle, prick them out into boxes and harden them off. Transfer the plants to their flowering positions during May, spacing them 15 inches (38cm) apart.

Sweet Basil may be used in all Italian tomato dishes and is the main ingredient in *pesto*.

O. minimum
BUSH BASIL

Hardy perennial

This herb, which seems to be a compact form of Sweet Basil, has aromatic leaves which are bright green above and grey-green on the underside. Small, white, tube-shaped flowers are produced during August. The plant rarely grows to more than 12 inches (30cm) in height. It requires a warm sheltered position.

The sowing instructions are the same as those for Sweet Basil and the herb is used in the same way.

Oenothera trichocalyx
EVENING PRIMROSE

Hardy biennial or half-hardy annual in a sunny position
Native to the Rocky Mountains of North America

This species has grey foliage and large white flowers which have a sweet scent. It grows to a height of 18 inches

(46cm). Any well-drained soil is suitable for this plant and it requires an open sunny position.

Sow the seeds February–March at a temperature of 50–60°F (10–15°C) and at a depth of ¼ inch (6mm). Germination takes 21–42 days. When the seedlings are large enough to handle, prick them out into boxes and harden them off. Transfer them to their flowering positions during May, spacing them 6 inches (15cm) apart. Alternatively, sow the seeds June–July in a nursery bed at a depth of ¼ inch (6mm). When the seedlings are large enough to handle, thin them out to 4 inches (10cm) apart and grow them on until the spring. Transfer the plants to their flowering positions during April, spacing them 6 inches (15cm) apart.

The flowers bloom during June.

The ornamental thistle, *Onopordum acanthium.*

Onopordum acanthium
ORNAMENTAL THISTLE

Hardy perennial
Native to Europe

This species of Thistle is covered with fine silver hairs and has large purple flowerheads. It grows to a height of 7 feet (2.1 metres). It is grown as a specimen plant or is used in large borders where it adds height to the

scheme. This plant grows well on any soil and in either a sunny or partially shaded position.

Sow the seeds May–June at a depth of ¼ inch (6mm) in a seed bed. Germination takes 21–42 days. When the seedlings are large enough to handle, thin them out to 12 inches (30cm) apart and grow them on until the autumn. Transfer the plants to their permanent positions during September, spacing them 30 inches (76cm) apart.

The flowers bloom July–August.

Orach, Red see *Atriplex hortensis rubra*

Origanum
MARJORAM

O. majorana
SWEET OR KNOTTED MARJORAM

Hardy annual
Native to Europe

This herb has bright-green leaves and clusters of white, mauve, or pink flowers. It grows to a height of 24 inches (61cm). It requires a well-drained soil and sunny position.

Sow the seeds February–March at a temperature of 50–60°F (10–15°C) and at a depth of ⅛ inch (3mm). When the seedlings are large enough to handle, plant them out, spacing them 12 inches (30cm) apart. Alternatively, sow the seeds April–May in open ground at a depth of ¼ inch (6mm). When the seedlings are large enough to handle, thin them out to 8–12 inches (20–30cm) apart.

The young shoots and leaves may be used in meat, fish, and tomato dishes, and also in salads.

O. onites
POT MARJORAM

Hardy perennial
Native to south-east Europe

This herb has bright-green aromatic leaves and mauve or white tube-

shaped flowers which appear
July–August. It grows to a height of 12
inches (30cm). It requires a rich soil
and sunny position.

The sowing instructions are the
same as those for *O. majorana*.

The leaves may be used to flavour
soups and stews.

Oswego Tea see *Monarda didyma*

Palm, Tie see *Cordyline indivisa*

Pansy see *Viola* × *wittrockiana*

Pansy, Tufted see *Viola* × *williamsii*

Papaver

POPPY

P. alpinum

ALPINE POPPY

Hardy perennial
Native to Europe

This miniature Poppy has grey-green
leaves and red, orange, yellow, or
white flowers. It grows to a height of
4–10 inches (10–25cm), depending on
the variety. It is suitable for use in rock
gardens and requires a well-drained
soil and sunny position.

Sow the seeds thinly June–August at
a depth of $\frac{1}{4}$ inch (6mm) in a well-
prepared seed bed. Germination takes
10–14 days. When the seedlings are
large enough to handle, thin them out
to 6 inches (15cm) apart. Transfer the
plants to their flowering positions
October–April, spacing them 8–12
inches (20–30cm) apart.

The flowers bloom May–July.

P. nudicaule

ICELAND POPPY

Hardy perennial
Native to Northern Europe

This large-flowered Poppy is available
in several pastel shades as well as red,
orange, yellow, and white. It grows to

Papaver nudicaule 'San Remo'.

a height of 18–30 inches (46–76cm),
depending on the variety. Like other
perennial Poppies, it requires a sunny
position and well-drained soil.

The sowing instructions are the
same as those for *P. alpinum*, but when
the plants are transferred to their
flowering positions, space them 12–18
inches (30–46cm) apart.

The flowers bloom June–July.

P. orientale

ORIENTAL POPPY

Hardy perennial
Native to Iran

The scarlet flowers of this species may
grow to 4 inches (10cm) in diameter.
The mid-green leaves are hairy.
Numerous garden varieties are
available in various colours including
pink, white, and orange, as well as
shades of red. Double-flowered forms
are also available. The plant grows to a
height of 24–36 inches (61–91cm).
Like other perennial Poppies, it
requires a sunny position and well-
drained soil.

The sowing instructions are the

same as those for *P. alpinum*, but when the plants are transferred to their flowering positions, space them 24 inches (61cm) apart.

The flowers bloom May–July.

Above: Papaver rhoeas, the Shirley poppy.

Below right: Papaver somniferum 'Paeony-Flowered Mixed'.

Below: A double-flowered variety of *Papaver rhoeas.*

P. rhoeas
FIELD POPPY, SHIRLEY POPPY

Hardy annual
Native to Europe

The foliage is pale green and the flowers are scarlet with black centres. The flowers may grow to 3 inches (8cm) in diameter. Cultivated varieties may be pink, white, or crimson. Double-flowered forms are available. The plant grows to a height of 24 inches (61cm). It requires a sunny position and well-drained soil.

Sow the seeds March–May at a depth of $\frac{1}{4}$ inch (6mm) in the flowering site. Germination takes 10–14 days. When the seedlings are large enough to handle thin them out to 12 inches (30cm) apart. Seeds may be sown in September to flower the following year.

The flowers bloom June–August.

P. somniferum
OPIUM POPPY

Hardy annual
Native to western Asia

This species has grey-green leaves and white, pink, scarlet, or purple flowers which may grow to 4 inches (10cm) in diameter. The plant grows to a height of 30–36 inches (76–91cm). Its large seedheads make it popular with

flower arrangers who use them for winter decorations. Double-flowered varieties are available. The plant requires a sunny position and well-drained soil.

The sowing instructions are the same as for *P. rhoeas*.

The flowers bloom June–August.

Parsley see *Petroselinum crispum crispum*

Parthenocissus tricuspidata 'Veitchii'

Ampelopsis veitchii
VIRGINIAN CREEPER

Hardy deciduous climber
Native to Japan

This plant has small yellow-green flowers which are followed by tiny dark-blue fruits. It grows to a height of 50 feet (15 metres). It is a rapid-growing climber which quickly covers sheds, garages, or houses. It is self-supporting, although in the early stages it may need to be tied to a cane or stick.

Sow the seeds during the spring in a cold frame or cool greenhouse at a depth of $\frac{1}{2}$ inch (1cm). Germination is sporadic, taking from 40 days onwards. When the seedlings are large enough to handle, prick them out singly into 3-inch (8-cm) pots and grow them on. Transfer the plants to their permanent positions September–October.

The flowers bloom June–July.

Passiflora caerulea

PASSION FLOWER

Half-hardy climber
Native to South America

This climbing plant has open star-shaped flowers which are white with purple coronas. It grows to a height of 15 feet (4.5 metres). It is usually grown

Passiflora caerulea: an attractive and unusual climbing plant.

on a trellis or against a wall, and requires a well-drained soil and sheltered position.

Sow the seeds February–March at a temperature of 68–78°F (20–25°C) and at a depth of $\frac{1}{4}$ inch (6mm). Germination takes 21–42 days. When the seedlings are large enough to handle, prick them out singly into 3-inch (8-cm) pots of potting compost and grow them on until the autumn. Transfer the plants to their permanent positions September–October.

The flowers bloom June–September.

Pelargonium

GERANIUM

Half-hardy perennial
Native to South Africa

P. × hortorum

This hybrid group of zonal Pelargoniums is the one most commonly grown. The leaves are mid-green and somewhat rounded in shape and have bronze or maroon markings. The flowers are available in many colours including white, pink, scarlet, crimson, and orange. The plants may

A selection of the many varieties of pelargonium. In an anti-clockwise direction from the top right they are 'Playboy' a dwarf F_1 hybrid; 'Mustang' a brilliant scarlet F_1 hybrid; 'Sprinter Mixture' an F_1 hybrid with colours ranging from deep-red, through salmon-pink, to white; 'Matador' from the Del Greco Series; and 'Stella' which is also from the Del Greco Series.

grow to a height of 6 feet (1.8 metres), but those most usually grown reach a height of 12–18 inches (30–46cm). Dwarf varieties are available and these grow to a height of 6–12 inches (15–30cm). Varieties grown for use in outdoor bedding schemes require a position in full sun and well-drained soil.

Sow the seeds December–January at a temperature of 68–78°F (20–25°C). They should be sown at a depth of $\frac{1}{4}$ inch (6mm) in pots or trays containing seed compost. Germination takes 14–28 days. When the seedlings are large enough to handle, prick them out singly into pots and grow them on at a temperature of 60–68°F (15–20°C). Harden them off carefully. Plant them out at the end of May or the beginning of June, spacing them about 12 inches (30cm) apart.

The flowers bloom May–October.

Penstemon
BEARD TONGUE

P. barbatus
Chelone barbata

Hardy perennial
Native to the southern USA

This species has long stems of tubular-shaped flowers in shades of pink,

cerise, and scarlet. It grows to a height of 36 inches (91cm). It requires a sunny position and well-drained soil.

Sow the seeds February–March at a temperature of 50–60°F (10–15°C) and at a depth of $\frac{1}{4}$ inch (6mm). Germination takes 14–28 days. When the seedlings are large enough to handle, prick them out into boxes and harden them off under a cold frame. Transfer the plants to their permanent positions during May, spacing them 24 inches (61cm) apart. Alternatively, sow the seeds June–July at a depth of $\frac{1}{4}$ inch (6mm) in the flowering site. When the seedlings are large enough to handle, thin them out to 24 inches (61cm) apart.

The flowers bloom June–August.

P. × gloxinioides

Hardy perennial
Native to the USA

Penstemon has large stems of snapdragonlike flowers in shades of scarlet or crimson. The flowers sometimes have white throats. The plant grows to a height of 24 inches (61cm). Any ordinary garden soil is suitable for this plant and although a

Two varieties of penstemon: 'Suttons Large-Flowered Mixed' (*left*) and 'Bouquet Mixture' (*below*).

sunny position is best, it is tolerant of partial shade.

Sow the seeds February–March at a temperature of 60–68°F (15–20°C) and at a depth of ⅛ inch (3mm). Germination takes 14–28 days. When the seedlings are large enough to handle, prick them out into boxes and grow them on. Harden them off for 2–3 weeks. Plant them out in the flowering site during August, spacing them 12–18 inches (30–46cm) apart. Alternatively, for flowering the following year, sow the seeds May–August at a depth of ¼ inch (6mm) in their flowering positions. When the seedlings are large enough to handle, thin them out to 12–18 inches (30–46cm) apart.

The flowers bloom June–July.

Pepper, Ornamental see *Capsicum annuum*

Perilla frutescens var nankinense

Half-hardy annual
Native to China

This plant has deeply indented, finely cut leaves which are bronze-purple, and spikes of small white flowers. It grows to a height of 24 inches (61cm). Its decorative foliage makes it an ideal plant for outdoor bedding schemes. It requires a sunny position and well-drained soil.

Sow the seeds February–March at a temperature of 68–78°F (20–25°C) and at a depth of ⅛ inch (3mm). Germination takes 14–28 days. When the seedlings are large enough to handle, prick them out into boxes and harden them off. Transfer the plants to their flowering positions during May, spacing them 12 inches (30cm) apart.

The flowers bloom July–August.

Periwinkle, Madagascar see *Vinca rosea*

Petroselinum crispum crispum
PARSLEY

Hardy biennial usually grown as an annual
Native to central and southern Europe

There are both plain and curly-leaved varieties of Parsley. The leaves are mid-green, and green-yellow flowers appear during the second summer. The plant grows to a height of 12–24 inches (30–61cm). Parsley requires a well-drained fertile soil which should be prepared by working in well-decayed manure or compost. It grows in either sun or partial shade.

Sow the seeds March–June at a depth of ¼ inch (6mm) in the open ground. When the seedlings are large enough to handle, thin them out to 3 inches (8cm) apart. Later, thin them out again to their final spacing of 9 inches (23cm) apart. Parsley may also be grown indoors in pots containing potting compost. Sprinkle 5–6 seeds on potting compost in a 4–6 inch (10–15cm) pot. Moisten the compost and place the pot in bright indirect light until germination has taken place. Then move it to a window sill or other position where it will be in full sun. Parsley is notoriously slow to germinate and the process may be speeded up if the drills in which the seeds are to be sown are watered with hot water before the seeds are sown.

Parsley has a mildly spicy flavour and contains the vitamins A and C. It has numerous culinary uses: it is an essential ingredient in a *bouquet garni* and in *maître d'hôtel* butter, and may be used in sauces and salad dressings.

Petunia × hybrida

Half-hardy perennial usually grown as a half-hardy annual
Native to Argentina

This is the garden hybrid of the Petunia family. There are numerous

There are numerous varieties of the garden petunia, *Petunia* x *hyrida*. They are available in a wide range of self and bicolor shades.

Top: 'Petunia Delight Mixture', an F$_1$ hybrid Double Multiflora.
Centre left: 'Razzle Dazzle', an F$_1$ hybrid Grandiflora.
Centre right: 'Star Joy', an F$_1$ hybrid Multiflora.
Right: 'Double Super Fanfare Mixed', an F$_1$ hybrid Double Grandiflora.

varieties which fall into four main groups: Multiflora, which grows to a height of 6–12 inches (15–30cm); Grandiflora, which grows to the same height as Multiflora; Nana Compacta, which grows to a height of 6 inches (15cm); and Pendula, which contains the trailing varieties of Petunia, and grows to a height of 6 inches (15cm). The foliage ranges from mid- to dark-green, and the trumpet-shaped flowers, which may grow to 5 inches (13cm) in diameter, are available in many colours including purple, scarlet, salmon-pink, and white. Single- and double-flowered forms are available. Petunias prefer light well-drained soil. Plant them in a sunny sheltered position as they do not tolerate cold, wet, or shady conditions.

Sow the seeds January–March at a temperature of 60–69°F (15–20°C) just pressing them into the surface of the seed compost. Germination takes 7–14 days. When the seedlings are large enough to handle, prick them out into boxes and grow them on. Harden them off, then transfer them to the flowering site during May, spacing them 12 inches (30cm) apart.

The flowers bloom from June until the first autumn frosts.

Phacelia campanularia

Hardy annual
Native to California

This plant has grey foliage and blue bell-shaped flowers. It grows to a height of 9 inches (23cm). It grows on most well-drained soils, but does best on sandy ones. Phacelia is best used in bold patches as an edging plant or in the rock garden. It may be used also as an early spring-flowering pot plant in the greenhouse.

Sow the seeds March–May in the flowering site at a depth of $\frac{1}{4}$ inch (6mm). Germination takes 10–21 days. When the seedlings are large enough

Phacelia campanularia.

to handle, thin them out to 6 inches (15cm) apart. The plants will flower about six weeks after sowing.

The flowers bloom June–September.

Pharbitis

P. purpurea
CONVOLVULUS MAJOR

Hardy annual
Native to tropical America

This plant has large funnel-shaped flowers in many colours. It grows to a height of 8 feet (2.4 metres). This is a rapid climber which quickly covers a fence or rambles up and through a hedge.

Sow the seeds March–May in open ground at a depth of $\frac{1}{2}$ inch (1cm). Germination takes 10–14 days.

The flowers bloom July–September.

P. tricolor
Ipomoea rubro-caeruleum
MORNING GLORY

Half-hardy annual
Native to tropical America

A climbing plant with large trumpet-shaped flowers which are blue, red-

Pharbitis tricolor 'Heavenly Blue'.

purple, or purple. It grows to a height of 8–10 feet (2.4–3 metres). If grown in a 6-inch (15-cm) pot, Pharbitis may be used as a display plant in a greenhouse. It also flourishes on well-cultivated soil in a sunny sheltered position.

Before sowing the seeds, soak them in water for 24 hours as the seed coats are very hard.

Sow 2–3 seeds March–April ½ inch (1cm) deep in a small pot at a temperature of 60–68°F (15–20°C). Germination takes 10–21 days. Transfer the plants to the flowering site during June. Alternatively, sow the seeds April–May in the flowering site at a depth of ½ inch (1cm). When the seedlings are large enough to handle, thin them out to 12 inches (30cm) apart.

The flowers bloom July–September.

Pheasant's Eye see *Adonis aestivalis*

Philodendron bipinnatifidum

Greenhouse evergreen
Native to South America

The dark-green leaves are deeply lobed. The plant grows to a height of 48 inches (122cm). It thrives in a position where it is shaded from direct sunlight.

Sow the seeds March–April at a temperature of 68–78°F (20–25°C) in a moist, soil-based seed compost. Just press the seeds into the surface of the compost, then lightly water it and cover it with polythene until germination takes place. Germination takes 14–21 days. During this period do not allow the soil to dry out. When the seedlings are large enough to handle, prick them out singly into 3-inch (8-cm) pots of potting compost. Pot on as necessary.

Phlox drummondii

Half-hardy annual
Native to Texas

These plants have light-green foliage and densely packed heads of flowers which may grow to 3 inches (8cm) in diameter. There are three main groups: Stellaris (Cuspidata), which has star-shaped flowers and grows to a height of 6 inches (15cm); Grandiflora, which has large flowers and grows to a height of 12 inches (30cm); and Nana

Phlox drummondii is one of the most popular of all the annuals. This variety is 'Suttons Beauty Mixed'.

Compacta, a dwarf strain with a bushy habit, which grows to a height of 9 inches (23cm). Phlox may be grown in window-boxes as well as in beds and borders. It requires a position in full sunshine and a well-drained soil.

Sow the seeds February–April at a temperature of 50–60°F (10–15°C) and at a depth of $\frac{1}{4}$ inch (6mm). Germination takes 10–14 days. When the seedlings are large enough to handle, prick them out into boxes, spacing them about 2 inches (5cm) apart. Harden them off, and then plant them out in their flowering positions during May, spacing them about 6–9 inches (15–23cm) apart. Seeds may also be sown September–October in a cold frame or greenhouse at a depth of $\frac{1}{4}$ inch (6mm) for flowering the following spring.

The flowers bloom June–August.

Physalis alkekengi franchetii

CHINESE LANTERN

Hardy perennial
Native to Japan

This plant has white flowers which are followed by orange bell-shaped fruits. It grows to a height of 24 inches (61cm). Any well-drained garden soil is suitable, but a sunny position is essential. This plant is popular with flower arrangers as it may be dried for use in winter decorations.

Sow the seeds February–March at a temperature of 50–60°F (10–15°C) and at a depth of $\frac{1}{4}$ inch (6mm). The seeds may also be sown May–July in a cold frame or cool greenhouse. Germination takes 14–28 days. When the seedlings are large enough to handle, prick them out into boxes and grow them on until the autumn. Transfer the plants to their permanent positions September–October, spacing them 24 inches (61cm) apart.

The flowers bloom July–August.

Pimpinella anisum

ANISEED

Hardy annual
Native to southern Europe, Egypt, and the Near East

The feathery foliage is brilliant green and the flowers are white. It grows to a height of 18 inches (46cm). It requires a sunny position and its seeds will not ripen in Britain unless the summer temperature is above-average.

Sow the seeds in shallow drills at a depth of $\frac{1}{4}$ inch (6mm) during April. Germination takes 10–21 days. When the seedlings are large enough to handle, space them out 12 inches (30cm) apart.

The seeds are used in confectionery and, when crushed and sprinkled on meat, enhance the flavour.

Pincushion Plant see *Cotula barbata*

Pink see *Dianthus plumarius*

Pink, Clove see *Dianthus caryophyllus*

Pink, Indian see *Dianthus chinensis*

Pink, Japanese see *Dianthus heddewigii*

Pink, Maiden see *Dianthus deltoides*

Pink, Sea see *Armeria maritima*

Poached Egg Plant see *Limnanthes douglasii*

Polka Dot Plant see *Hypoestes sanguinolenta*

Polyanthus see *Primula polyantha*

Pomegranate, Dwarf see *Punica granatum nanum*

Poor Man's Orchid see *Schizanthus pinnatus*

Poppy see *Papaver*

Poppy, Alpine see *Papaver alpinum*

Poppy, Californian see *Eschscholzia californica*

Poppy, Field see *Papaver rhoeas*

Poppy, Himalayan Blue see *Meconopsis betonicifolia*

Poppy, Iceland see *Papaver nudicaule*

Poppy, Opium see *Papaver somniferum*

Poppy, Oriental see *Papaver orientale*

Poppy, Shirley see *Papaver rhoeas*

Poppy, Welsh see *Meconopsis cambrica*

Portulaca

P. grandiflora

Half-hardy annual
Native to Brazil

This species has cup-shaped flowers which are brightly coloured in a wide range of shades including orange, scarlet, purple, and white. It grows to a height of 6–9 inches (15–23cm) and is useful as a ground-cover plant. It requires a sunny position and well-drained soil.

Portulaca grandiflora: this variety, 'Suttons Improved Double Mixed' has semi-double flowers.

Sow the seeds February–March at a temperature of 60–68°F (15–20°C) and at a depth of $\frac{1}{8}$ inch (3mm). Germination takes 14–21 days. When the seedlings are large enough to handle, prick them out into boxes and harden them off. Transfer the plants to their flowering positions during May, spacing them 6 inches (15cm) apart. Alternatively, sow the seeds April–May in the flowering site at a depth of $\frac{1}{4}$ inch (6mm). When the seedlings are large enough to handle, thin them out 6 inches (15cm) apart.

The flowers bloom June–September.

P. oleracea

PURSLANE

Half-hardy annual
Native to Europe

Purslane has propeller-like leaves and crimson stems. It grows to a height of 4 inches (10cm). It is tolerant of dry conditions and requires an open sunny position and well-drained soil.

Sow the seeds in May in shallow drills at a depth of $\frac{1}{4}$ inch (6mm) where the plants are to grow. Germination takes 14–21 days. When the seedlings are large enough to handle, space them out 6 inches (15cm) apart.

The leaves may be used raw in salads, or cooked and used as a vegetable.

Poterium sanguisorba see *Sanguisorba minor*

Primrose see *Primula vulgaris*

Primrose, Cape see *Streptocarpus*

Primrose, Evening see *Oenothera trichocalyx*

Primrose, Fairy see *Primula malacoides*

Primrose, Japanese see *Primula japonica*

Primula

Greenhouse species

P. × kewensis

Greenhouse perennial
Native to China

This species has fragrant and long-lasting buttercup-yellow flowers. It grows to a height of 8 inches (20cm).

For flowers the following spring, sow the seeds May–July in a cold frame or cool greenhouse. Barely cover the seeds with compost. Germination takes 10–21 days. When the seedlings are large enough to handle, prick them out singly into pots of potting compost. Pot on as necessary but do not overpot. A $3\frac{1}{2}$-inch (9-cm) pot is adequate for final potting.

The flowers bloom December–April.

Primula malacoides: two varieties from Suttons Large-Flowered Strain.

Primula kewensis.

P. malacoides

FAIRY PRIMROSE

Greenhouse perennial usually grown as a greenhouse annual
Native to China

This species of Primula has pale-green foliage and starlike flowers which range in colour from lilac, through rose-pink, to white. It grows to a height of 8 inches (20cm).

Sow the seeds May–August in a cold frame or greenhouse, barely covering the seeds with compost. Germination takes 10–21 days. When the seedlings are large enough to handle, prick them

out into boxes and grow them on.
When the seedlings are well
developed, transplant each one to a
3-inch (8-cm) pot of potting compost.
Pot on as necessary.

The flowers bloom late December–
April.

P. obconica

Greenhouse perennial usually grown
as an annual
Native to China

This plant has light-green, slightly
hairy leaves, and clusters of red,
apricot, blue, pink, white, or violet
flowers. It grows to a height of 9–15
inches (23–38cm), depending on the
variety. N.B. The leaves may cause an
allergic skin reaction in some people.
It requires a cool situation in the house
or greenhouse and plenty of light and
air. However, keep it shaded from
direct sunlight during the summer.

Sow the seeds February–June at a
temperature of 60–68°F (15–20°C) and
at a depth of $\frac{1}{8}$ inch (3mm).
Germination takes 10–21 days. When
the seedlings are large enough to
handle, prick them out into small pots
containing potting compost. Pot on as
necessary until the plants are in 5-inch
(13-cm) pots. Keep the soil moist, but
do not overwater the plants.

The flowers bloom December–May.

P. sinensis

Greenhouse perennial usually grown
as a greenhouse annual
Native to China

This species has bright-green hairy
leaves with the flowers carried in
clusters of three. It grows to a height
of 10 inches (25cm).

The sowing instructions are the
same as those for P. × kewensis.

The flowers bloom December–
March.

Above: Primula obconica 'La Scala'.
Below: Primula obconica 'Giant Mixed' from
Suttons Giant-Flowered Strain.

Hardy species

P. denticulata

DRUMSTICK PRIMULA

Hardy perennial
Native to the Himalayas

This species has pale-green foliage and
mauve flowers. It grows to a height of
12 inches (30cm). It requires a sunny

or partially shaded position and fertile soil which does not dry out during the spring and summer months. It is suitable for use in spring bedding schemes and for rock gardens.

Sow the seeds March–May in a cold frame or cool greenhouse, barely covering them with compost. Germination takes 21–42 days. When the seedlings are large enough to handle, prick them out into boxes and grow them on until the early autumn. Transfer the plants to their permanent positions during September, spacing them 9 inches (23cm) apart.

The flowers bloom March–May.

P. florindae
GIANT COWSLIP

Hardy perennial
Native to Tibet

This species has mid-green leaves and yellow trumpet-shaped flowers. It grows to a height of 36 inches (91cm). It thrives on moist soil.

The sowing instructions are the same as those for *P. denticulata*, but when the plants are transferred to their permanent positions, space them 12–15 inches (30–38cm) apart.

The flowers bloom June–July.

Right: Primula polyantha 'Lemon Punch' an F$_1$ hybrid.

Below: Primula florindae.

P. japonica
JAPANESE PRIMROSE

Hardy perennial
Native to Japan

The colours of this species' flowers range from white to crimson. It grows to a height of about 18 inches (46cm). Like *P. florindae*, it thrives on moist soil.

The sowing instructions are the same as those for *P. denticulata*, but when the plants are transferred to their permanent positions, space them 9–12 inches (23–30cm) apart.

The flowers bloom May–July.

P. polyantha
POLYANTHUS

Hardy perennial
Native to Europe

This garden hybrid has been bred from *P. vulgaris* (Primrose) and *P. veris* (Cowslip). Polyanthus flowers are brilliantly coloured and are available in blue, cream, yellow, white, pink, crimson, and scarlet, all with yellow centres. The plant grows to a height of 8–10 inches (20–25cm). It thrives on any well-drained soil and is suitable for use in sheltered borders.

Primula polyantha 'Spring Promise' an F₁ hybrid.

Sow the seeds January–March at a temperature of 60–68°F (15–20°C) and at a depth of ⅛ inch (3mm). Germination takes 21–50 days. When the seedlings are large enough to handle, prick them out into pots and grow them on. Plant them out in their flowering positions during September, spacing them about 8 inches (20cm) apart. Alternatively, sow the seeds outdoors in a seed bed during May. When the plants are large enough to handle, thin them out to 6 inches (15cm) apart.

The flowers bloom during the spring.

P. × pubescens
AURICULA

Hardy perennial
Native to Europe

This plant forms clusters of grey-green leaves and has yellow or purple flowers. It grows to a height of 6 inches (15cm). It is suitable for open borders, as it survives the hardest winter, and also for rockeries.

Sow the seeds January–April at a temperature of 50–60°F (10–15°C) and at a depth of ⅛ inch (3mm). Germination takes 21–35 days. When the seedlings are large enough to handle, prick them out into boxes and grow them on. Transfer the plants to their permanent positions during the autumn, spacing them 9 inches (23cm) apart.

The flowers bloom March–May.

P. rosea

Hardy perennial
Native to the Himalayas

This species has mid-green leaves and shocking-pink flowers. It grows to a height of 6 inches (15cm). For the best results, grow it on very moist soil. It is ideal as a plant for the water's edge.

The sowing instructions are the same as those for *P. denticulata*, but when the plants are transferred to their permanent positions, space them far enough apart to allow for their spread of about 8 inches (20cm).

The flowers bloom March–April.

P. veris
COWSLIP

Hardy perennial
Native to Europe

This cultivated variety of the wild flower is available in many colours including red, yellow, and crimson. It grows to a height of 6 inches (15cm) and succeeds in both sunny and shady positions.

Sow the seeds March–July at a depth of ⅛ inch (3mm) in boxes of sterilized compost. Keep the temperature at 50–60°F (10–15°C). Germination takes 28–42 days. When the seedlings are large enough to handle, prick them out into boxes and grow them on until the autumn. Transfer the plants to their flowering positions during September, spacing them 9 inches (23cm) apart.

The flowers bloom during the spring.

P. vulgaris
PRIMROSE

Hardy perennial
Native to Europe

The garden forms of this species are available in many colours including orange, yellow, white, blue, and salmon-pink. It grows to a height of 4 inches (10cm). Primroses are very easy to grow and succeed in both sunny and shady positions.

Two garden forms of *Primula vulgaris. Above:* 'Juliet Mixed' a miniature Primrose; (*below*): 'Colour Magic' an F₁ hybrid.

Sow the seeds January–March at a temperature of 60–68°F (15–20°C) and at a depth of $\frac{1}{8}$ inch (3mm). Germination takes 21–42 days. When the seedlings are large enough to handle, prick them out into boxes and grow them on until the autumn. Transfer the plants to their permanent positions September–October, spacing them 9 inches (23cm) apart. Alternatively, sow the seeds May–June in a seed bed at a depth of $\frac{1}{4}$ inch (6mm). When the seedlings are large enough to handle, thin them out to 6 inches (15cm) apart and grow them on until the autumn. Transfer the plants to their permanent positions during September, spacing them 9 inches (23cm) apart.

The flowers bloom March–April.

Primula, Drumstick see *Primula denticulata*

Prince's Feather see *Amaranthus hypochondriacus*

Prince of Wales' Feather see *Celosia argentea plumosa*

Punica granatum nanum
DWARF POMEGRANATE

Greenhouse perennial
Native to western Asia

The green leaves of this plant are small and shiny. The orange flowers, which bloom during the summer, are followed by tiny pomegranates. The plant grows about 7 inches (18cm) during its first year, and achieves a final height of about 18 inches (46cm). It is grown as a pot plant in the greenhouse or as a houseplant.

Sow the seeds March–May at a temperature of 60–68°F (15–20°C) and at a depth of $\frac{1}{4}$ inch (6mm). Germination takes 21–42 days. When the seedlings are large enough to handle, prick them out into boxes and grow them on. When the seedlings are well developed, transplant each one to a 3-inch (8-cm) pot of potting compost. Pot on as necessary.

The flowers bloom June–September.

Purslane see *Portulaca oleracea*

Purslane, Rock see *Calandrinia umbellata*

Pygmy Torch see *Amaranthus hypochondriacus*

Pyrethrum roseum see *Chrysanthemum coccineum*

Queen Anne's Lace see *Didiscus caerulus*

Quince, Japanese see *Chaenomeles japonica*

Quince, Maule's see *Chaenomeles japonica*

Ranunculus aconitifolius
FAIR MAIDS OF FRANCE

Hardy perennial
Native to Europe

This plant has mid-green leaves and white flowers which are $\frac{1}{2}$ inch (1cm) across. It grows to a height of 12 inches (30cm). For best results, give it either a sunny or partially shaded position. It requires a moist soil and is suitable for use as a border plant.

Sow the seeds March–June in a cold frame or cool greenhouse at a depth of $\frac{1}{8}$ inch (3mm). Germination takes 21–42 days. When the seedlings are large enough to handle, prick them out into boxes and grow them on until the autumn. Transfer the plants to their permanent positions September–October, spacing them 12–18 inches (30–45cm) apart.

The flowers bloom May–June.

Reseda odorata 'Sweet-scented'.

Rechsteineria cardinalis
Greenhouse perennial
Native to South America

This pot plant has velvety green leaves and spikes of scarlet tube-shaped flowers. It grows to a height of 8 inches (20cm).

Sow the seeds January–April at a temperature of 68–78°F (20–25°C). Just press the seeds into the surface of the seed compost. Germination takes 21–28 days. When the seedlings are large enough to handle, prick them out singly into 3-inch (8-cm) pots of potting compost.

The flowers bloom June–August.

Red Hot Poker see *Kniphofia*

Red Mountain Spinach see *Atriplex hortensis rubra*

Reseda odorata
MIGNONETTE

Hardy annual
Native to Egypt

This cottage-garden flower has a distinctive scent which is very attractive to bees. It has mid-green foliage and small yellow-white flowers

which are carried on loose heads. It grows to a height of 12–30 inches (30–76cm). This plant thrives in either a sunny or partially shaded position but it requires a well-prepared soil. It germinates more freely if the soil is firmed down before the seeds are sown. If the soil is acid, it is advisable to add some lime.

Sow the seeds March–May in the flowering site at a depth of $\frac{1}{8}$ inch (3mm). Germination takes 10–14 days. When the seedlings are large enough to handle, thin them out to 9–12 inches (23–30cm) apart. Seeds sown in a greenhouse August–September will flower early the following spring.

The flowers bloom June–October.

Rhodanthe manglesii see
Helipterum manglesii

Ricinus communis
CASTOR-OIL PLANT

Half-hardy annual
Native to tropical Africa

This plant has mid-green or dark-purple palmlike leaves which may grow to 12 inches (30cm) in diameter. It grows to a height of 48 inches (122cm). Although this plant is usually grown as a pot plant either in the greenhouse or indoors, it may be used in the garden to give beds and borders a tropical appearance.

Sow the seeds January–March at a temperature of 60–68°F (15–20°C) and at a depth of $\frac{1}{2}$ inch (1cm). Germination takes 14–21 days. When the seedlings are large enough to handle, prick out those that are to be used as pot plants and plant each one in a 5-inch (13-cm) pot of potting compost. Prick out those that are to be planted outside into boxes and harden them off before planting them out during May. Alternatively, sow the seeds in the flowering site at a depth of

Ricinus communis 'Gibsoni'.

$\frac{1}{2}$ inch (1cm) during May. Germination takes 14–21 days. When the seedlings are large enough to handle, thin them out to 36 inches (91cm) apart.

The insignificant green flowers bloom during July.

Rock Cistus see *Helianthemum*

Rosa polyantha nana
FAIRY ROSE

Hardy perennial
Native to Europe

This tiny Rose tree is covered with single or semi-double flowers which are white or pale pink. It grows to a height of 12 inches (30cm) and is ideal for growing in pots or window-boxes. It is also suitable for small beds in the garden. If it is to be grown out of doors it requires a sunny position and well-drained soil.

Sow the seeds April–July at a depth of $\frac{1}{2}$ inch (1cm) in a cold frame or cool greenhouse. Germination is sporadic, taking from 50 days onwards. When the seedlings are large enough to handle, prick them out into pots of potting compost, allowing several seedlings to each 5-inch (13-cm) pot.

The flowers bloom June–July.

Rose of Sharon see *Hypericum calycinum*

Rose, Sun see *Helianthemum*

Rubber Plant see *Ficus elastica*

Rudbeckia hirta
CONEFLOWER, BLACK-EYED SUSAN

Hardy annual
Native to the USA

The daisylike flowers have a central cone-shaped disc which contrasts with the colour of the surrounding petals. The colours range from lemon-yellow to bronze. The plant grows to a height of 18–36 inches (46–91cm). It requires a well-drained soil and sunny position.

Sow the seeds February–March at a temperature of 60–68°F (15–20°C) and at a depth of $\frac{1}{8}$ inch (3mm). Germination takes 14–21 days. When the seedlings are large enough to handle, prick them out into boxes and grow them on. Harden them off and then transfer them to their flowering positions during May, spacing them 12–18 inches (30–46cm) apart. Alternatively, sow the seeds April–May at a depth of $\frac{1}{4}$ inch (6mm) in the flowering site. Germination takes 14–21 days. When the seedlings are large enough to handle, thin them out to 12–18 inches (30–46cm) apart.

The flowers bloom August–October.

Rue see *Ruta graveolens*

Rumex scutatus
FRENCH SORREL

Hardy perennial
Native to Europe and Asia

Light-green leaves shaped like arrow heads form dense clusters of foliage; red-green flowers appear during the spring. The plant grows to a height of 12–18 inches (30–46cm). Sorrel should be grown on a rich moist soil

Rudbeckia hirta: 'Marmalade' (*above*) and 'Rustic Dwarfs' (*below*).

Rudbeckia hirta 'Irish Eyes' is unique in having an emerald-green eye.

and, although it does best in a sunny position, it also grows in partial shade.

Sow the seeds during April in a seed bed at a depth of $\frac{1}{4}$ inch (6mm). When the seedlings are large enough to handle, (ie, when they are about 1–2 inches (2.5–5cm) tall), transplant them to their permanent positions, spacing them 6–8 inches (15–20cm) apart.

The young leaves may be used in salads, to which they add a lemony flavour; older leaves may be cooked and eaten like spinach.

Saintpaulia ionantha is one of the most popular of all houseplants.

Ruta graveolens

RUE

Hardy perennial evergreen
Native to southern Europe

Rue has deep blue-green leaves and pale-yellow flowers which appear June–July. The plant grows to a height of 24–36 inches (61–91cm). It requires a well-drained soil and sunny position. This herb is now grown mainly as a decorative foliage plant.

Sow the seeds during the early summer in a shallow drill in a seed bed. When the seedlings are large enough to handle, thin them out to 12 inches (30cm) apart. Transplant them to their permanent positions during September, spacing them 18 inches (46cm) apart.

The leaves may be finely chopped and added sparingly to salads, to which they impart a bitter flavour.

Sage see *Salvia officinalis*

St John's Wort see *Hypericum calycinum*

Saintpaulia ionantha

AFRICAN VIOLET

Greenhouse perennial
Native to Tanzania

The leaves have a velvety texture and the flowers are white, blue-purple, or rose-pink. The plant grows to a height of 4 inches (10cm). This is a popular houseplant which thrives in a moist atmosphere. *N.B.* If splashed with water the leaves will mark.

Sow the seeds January–March at a temperature of 68–78°F (20–25°C). Just press the seeds into the surface of the seed compost. Germination takes 21–42 days. When the seedlings are large enough to handle, prick them out into boxes and grow them on. When the seedlings are well developed, prick them out singly into 3-inch (8-cm) pots of potting compost. Pot on as necessary. Keep the compost moist, but do not allow it to become soggy.

The flowers bloom almost all the year, but they bloom most freely June–October.

Salpiglossis sinuata

Half-hardy annual
Native to Argentina

The very ornamental flowers are veined and funnel-shaped. Colours available include crimson, lavender, orange, yellow, and scarlet. The plant grows to a height of 24 inches (61cm). It requires an open sunny position and moderately rich soil.

Sow the seeds January–March at a

temperature of 60–68°F (15–20°C) and at a depth of ⅛ inch (3mm). Germination takes 10–14 days. When the seedlings are large enough to handle, prick them out into boxes and harden them off. Transfer the plants to their flowering positions during May spacing them 12 inches (30cm) apart. Alternatively, sow the seeds April–May in the flowering site at a depth of ¼ inch (6mm). When the seedlings are large enough to handle, thin them out to 12 inches (30cm) apart.

The flowers bloom July–September.

Salvia

S. horminum

Hardy annual
Native to southern Europe

This plant is grown mainly for its attractive coloured bracts which are purplish-blue, white, rose or red. The flowers are pale pink or purple. The plant grows to a height of 18 inches (46cm) and requires an open, sunny position and a well-drained soil. It is useful for flower arrangers as it lasts

Above and below: two varieties of *Salpiglossis sinuata.*

Two scarlet-flowered varieties of *Salvia horminum,* 'Carabiniere' (*above*) and 'Hot Shot' (*below*).

Left: Salvia horminum 'Bouquet Mixed'.
Below: Salvia officinalis.

well in water when cut and may be dried for winter decorations.

Sow the seeds March–May at a depth of $\frac{1}{4}$ inch (6mm) in the open ground where they are to flower. Germination takes 14–21 days. When the seedlings are large enough to handle, thin them out to 9 inches (23cm) apart.

The flowers bloom July–September.

S. officinalis

SAGE

Hardy perennial
Native to southern Europe

Sage has grey-green leaves, and violet-blue flowers which appear in June and July. It grows to a height of 24 inches (61cm). It requires a light well-drained soil.

Sow the seeds February–March at a temperature of 60–68°F (15–20°C) in trays or pots of seed compost. Alternatively, sow the seeds April–May in shallow drills in a seed bed. Germination takes 21–35 days. When the seedlings are large enough to handle, transplant them to a nursery bed, spacing them about 6 inches (15cm) apart. Later, transfer them to

their permanent positions, spacing them 15 inches (38cm) apart.

Sage leaves dry well. It is traditionally used with fatty meats such as pork, duck, and goose. It may be used also with cheese, onions, and liver.

S. patens

Half-hardy perennial usually grown as a half-hardy annual
Native to Mexico

A compact species of Salvia with pointed mid-green leaves and tube-shaped brilliant-blue flowers. It grows to a height of 24 inches (61cm). This plant requires an open sunny position and a well-drained soil.

Sow the seeds January–March at a temperature of 68–78°F (20–25°C) and at a depth of $\frac{1}{4}$ inch (6mm). Germination takes 14–21 days. When the seedlings are large enough to handle, prick them out into boxes and harden them off. Plant them out in their flowering positions from late May onwards, spacing them about 10 inches (25cm) apart.

The flowers bloom August–September.

S. sclarea
CLARY

Hardy biennial usually grown as an annual
Native to the Mediterranean region

The leaves are large and hairy, and the flowers, which appear in August, are pinkish-mauve with deep pink bracts. The plant grows to a height of about 30 inches (76cm). Ideally, it requires a sunny position and a fertile well-drained soil.

Sow the seeds during April in the open ground at a depth of $\frac{1}{4}$ inch (6mm). Germination takes 14–21 days. When the seedlings are large enough to handle, thin them out to 12 inches (30cm) apart.

An infusion of Clary leaves was an old-fashioned remedy for sore eyes—its name is an abbreviation of 'clear eyes'. The leaves may be fried in batter and served with lemon juice as an accompaniment to omelettes and meat dishes. Although the leaves are usually used fresh, they may be dried or frozen.

S. splendens

Half-hardy perennial usually grown as a half-hardy annual
Native to Mexico

This species has bright-green foliage, and scarlet flowers which are carried on broad spikes. It reaches a height of 12–15 inches (30–38cm). It requires an open sunny position and well-drained soil.

The sowing instructions are the same as those for S. patens.

The flowers bloom from July until the first autumn frosts.

Sandwort see *Arenaria montana*

Sanguisorba minor
Poterium sanguisorba
BURNET

Hardy perennial
Native to Europe

The leaves resemble those of the Wild Rose: tufts of greenish flowers with purple-red stamens appear during the early summer. Burnet grows to a height of 6 inches (15cm). It grows best on a well-drained soil and requires a sunny position.

Sow the seeds in the open ground at a depth of $\frac{1}{2}$ inch (1cm) during April. Germination takes 14–35 days. When the seedlings are large enough to handle, thin them out to 12 inches (30cm) apart.

The leaves, which have a delicate cucumber flavour, may be used in salads and to flavour drinks.

Saponaria

S. ocymoides

Hardy perennial
Native to Europe

This species has pale-green leaves, and rose-coloured star-shaped flowers which are so numerous that they hide the foliage. It grows to a height of 3 inches (8cm). This trailing plant is suitable for rockeries and may be grown also on walls. It thrives on any fertile garden soil but it requires a sunny position.

Sow the seeds March–May at a depth of $\frac{1}{4}$ inch (6mm) in the flowering site. Germination takes 10–14 days. When the seedlings are large enough to handle, thin them out to 12 inches (30cm) apart.

The flowers bloom May–July.

S. vaccaria
Vaccaria pyramidata

Hardy annual
Native to southern Europe

This species has light-green foliage and graceful sprays of dainty pink flowers. It grows to a height of 24 inches (61cm). It requires the same conditions as *S. ocymoides* and is suitable for planting in annual borders. It is very useful as a cut flower.

The sowing instructions are the same as those for *S. ocymoides*, but when the plants are thinned out, space them out 9 inches (23cm) apart. They will require staking.

The flowers bloom June–August.

Satureia
SAVORY

S. hortensis
SUMMER SAVORY

Hardy annual
Native to Europe

This herb has dark-green leaves, hairy stems, and lilac tube-like flowers which bloom July–September. The plant grows to a height of 12 inches (30cm). It requires a sunny position and a well-drained soil.

Sow the seeds during April in open ground at a depth of $\frac{1}{4}$ inch (6mm). Germination takes 14–21 days. When the seedlings are large enough to handle, thin them out to 6–9 inches (15–23cm) apart. To maintain a supply of Summer Savory throughout the year, sow a few seeds in a pot of potting compost in September and grow them in a greenhouse or indoors at a temperature of about 50°F (10°C).

The leaves of Summer Savory may be added to soups, and meat and fish dishes. They may be used also in pot-pourri.

S. monatana
WINTER SAVORY

Hardy perennial
Native to Europe and Asia

This herb has grey-green leaves and woody stems. The rose-purple tube-shaped flowers bloom July–October. The plant grows to a height of 12 inches (30cm). It requires a sunny position and well-drained soil.

Sow the seeds during April in the open ground at a depth of $\frac{1}{4}$ inch (6mm). Germination takes 14–21 days. When the seedlings are large enough to handle, thin them out to 9–12 inches (23–30cm) apart.

The leaves of Winter Savory may be used in the same dishes as those of Summer Savory, but they have a coarser flavour.

Saxifraga moschata
MOSSY SAXIFRAGE

Hardy perennial
Native to Europe

Mossy Saxifrage forms hummocks of bright-green foliage which produce clusters of small blooms in shades of yellow and white. Colours of cultivated varieties include rose, pink, and red. The plant grows to a height of 6 inches (15cm). It thrives on most soils, but prefers a sunny position. It is suitable for growing in rockeries.

Sow the seeds March–July in a greenhouse or cold frame at a depth of $\frac{1}{8}$ inch (3mm). Germination takes 21–42 days. When the seedlings are large enough to handle, prick them out into boxes and grow them on until the autumn. Transfer the plants to their permanent positions September–October, spacing them far enough apart to allow for their spread of up to 18 inches (46cm).

The flowers bloom April–May.

An autumn border containing dahlias, chrysanthemums, violas and rudbeckias.

Scabiosa
SCABIOUS

S. atropurpurea
SWEET SCABIOUS

Hardy annual
Native to Europe

Numerous cultivated varieties of this species are available from seedsmen. Colours include many shades of purple as well as blue, red, pink, and white. The plant grows to a height of 36 inches (91cm). Dwarf varieties are available and reach a height of 15–18 inches (38–46cm). Scabious requires a sunny position and a fertile well-drained soil.

S. caucasica
CAUCASIAN SCABIOUS

Hardy perennial
Native to the Caucasus

The garden varieties of this species have flowers in white or shades of blue or purple. The plant grows to a height of 18–24 inches (46–61cm). It requires a sunny position and a fertile well-drained soil.

Sow the seeds February–March at a temperature of 60–68°F (15–20°C) and at a depth of $\frac{1}{4}$ inch (6mm). Germination takes 14–21 days. When the seedlings are large enough to handle, prick them out into boxes and grow them on. Transfer the plants to the flowering site September–March, spacing them 18 inches (46cm) apart.

The flowers bloom June–September.

Above: A double-flowered variety of scabious
Right: Schefflera actinophylla.

Schefflera actinophylla

Greenhouse perennial
Native to Melanesia

This evergreen shrub has mid-green glossy leaves which develop from the stem in a raylike fashion. The leaves grow 3–6 inches (8–15cm) in length. The plant grows to a height of about 6 feet (1.8 metres). It is grown as a foliage plant in the greenhouse or as a houseplant.

Sow the seeds at any time at a

Sow the seeds April–May at a depth of $\frac{1}{2}$ inch (1cm) in the flowering site. Germination takes 14–21 days. When the seedlings are large enough to handle, thin them out to 9 inches (23cm) apart. Alternatively, sow the seeds at a depth of $\frac{1}{2}$ inch (1cm) in the flowering site during September for flowers the following spring.

The flowers bloom July–September.

temperature of 68–78°F (20–25°C). They should be sown at a depth of ⅛ inch (3mm) in a soil-based seed compost. Germination takes 14–21 days. When the seedlings are large enough to handle, prick them out singly into 2½-inch (6-cm) pots of potting compost. Pot on as necessary. Once they are well developed the seedlings will survive at much lower temperatures, but not lower than 40°F (5°C).

Schizanthus pinnatus
BUTTERFLY FLOWER, POOR MAN'S ORCHID

Half-hardy annual
Native to Chile

The pale-green leaves are fernlike and the flowers resemble Orchids. A wide variety of colours is available and many flowers are marked or spotted with contrasting colours. The plants grow to a height of 18–48 inches (46–122cm). These plants are usually grown in pots in a cool greenhouse, but they can be grown outside in the spring. If they are to be grown outside, they require a sunny sheltered position and a light well-drained soil.

Sow the seeds during August at a temperature of 60–68°F (15–20°C) and at a depth of ⅛ inch (3mm). Germination takes 7–14 days. As soon as the seedlings are large enough to handle, pot them up individually into 3-inch (8-cm) pots. Pot on as soon as the roots are seen at the edge of the soil ball. If they are not kept growing, they will come into flower prematurely. Continue potting until the plants are in 6-inch (15-cm), 7-inch (18-cm), or 8-inch (20-cm) pots, depending on how large a plant is required. Keep the plants frost-free but not too warm during the winter, and in the spring they will be a mass of flowers.

These plants may be flowered out of doors during the summer. Sow the

Two large-flowered varieties of *Schizanthus pinnatus*: 'Suttons Giant Hybrids Mixed' (*above*) and 'Hit Parade' (*below*).

Schizanthus pinnatus 'Star Parade', a dwarf variety.

seeds at a temperature of 60–68°F
(15–20°C) during March. When the
plants are large enough to handle,
prick them out and harden them off.
Plant them out in May, spacing them 9
inches (23cm) apart. Alternatively,
sow the seeds in the flowering site
during May and then thin them to 9
inches (23cm) apart.

Plants sown in August will flower
March–April; plants grown out of
doors will flower late June–September.

Senecio

S. bicolor
S. cineraria, S. maritimus, Cineraria
maritima

Half-hardy perennial
Native to the Mediterranean region

The foliage is covered with white
woolly hairs which give it a silvery
appearance. The flowers are yellow. It
grows to a height of 24 inches
(61cm), although dwarf forms are
available which grow to a height of 6–8
inches (15–20cm). Any ordinary

garden soil is suitable for this plant,
and its attractive foliage makes it
particularly useful for bedding
schemes.

Sow the seeds February–April at a
temperature of 60–68°F (15–20°C).
Use boxes or pots of seed compost and
sow the seeds at a depth of $\frac{1}{8}$ inch
(3mm). Germination takes 10–14 days.
When the seedlings are large enough
to handle, prick them out into boxes of
potting compost. Transfer the plants
to their permanent positions
May–early June, spacing them 12
inches (30cm) apart.

The flowers bloom July–September.

S. cruentus
Cineraria cruenta
CINERARIA

Greenhouse perennial
Native to the Canary Islands

The foliage varies from mid- to dark-
green, and the colours of the flowers
include white, mauve, red, blue, and
pink. Bicolor varieties are also
available. The plant grows to a height
of 10–18 inches (25–46cm), depending

Above: Senecio bicolor 'Silverdust' is an
excellent bedding plant adding a cool note to
brilliantly-coloured summer borders.

Left: Senecio bicolor 'Early Flowering Spring
Glory', a dwarf Cineraria.

on the variety. It grows best in a cool moist atmosphere, and must be protected from direct sunshine. The numerous varieties of this species are very popular as pot plants.

Sow the seeds April–August at a temperature of 50–60°F (10–15°C) for a succession of flowers throughout the winter months. The seeds should be sown in pots or boxes of seed compost at a depth of ⅛ inch (3mm). Germination takes 10–14 days. Water moderately and do not allow the soil to dry out. Pot on as necessary.

The flowers bloom December–April.

Silene schafta is an ideal edging plant.

S. elegans
JACOBEA

Hardy annual
Native to South Africa

This species has dark-green foliage and carmine-rose flowers which resemble Cinerarias. It grows to a height of 18 inches (46cm). It requires a sunny position and well-drained soil. Plant it in bold groups in the border for splashes of colour. If planted in an exposed position it may need staking.

Sow the seeds March–May at a depth of ¼ inch (6mm) in the flowering site. Germination takes 14–21 days. When the seedlings are large enough to handle, thin them out to 6 inches (15cm) apart.

The flowers bloom July–August.

Sensitive Plant see *Mimosa pudica*

Shoo Fly Plant see *Nicandra physaloides*

Silene schafta
CAMPION

Hardy perennial
Native to the Caucasus

This species has mid-green leaves and bears purplish-pink flowers. It grows to a height of 4–6 inches (10–15cm). It thrives on any well-drained soil and requires either a sunny or partially shaded position. It is suitable as a permanent edging for a bed or border and for the rockery.

Sow the seeds during April or June–September in the flowering site at a depth of ¼ inch (6mm). Germination takes 10–21 days. When the seedlings are large enough to handle, thin them out far enough apart to allow for their spread of up to 12 inches (30cm).

The flowers bloom July–October.

Silver Lace see *Chrysanthemum ptarmicaeflorum*

Silver Wattle see *Acacia dealbata*

Sinningia speciosa
GLOXINIA

Greenhouse perennial
Native to Brazil

This species has dark-green foliage. The trumpet-shaped flowers have velvety flowers in shades of red, pink, purple, and white. The plant grows to a height of 10 inches (25cm). In temperate climates it is suitable for the

Sinningia speciosa: a greenhouse plant with rich velvety-textured petals. *Above:* an example of the 'Triumph' Strain; *below:* 'Satin Beauty'.

Slipper Flower see *Calceolaria*

Smilax see *Asparagus medeoloides*

Smithiantha hybrids
TEMPLE BELLS

Greenhouse perennial
Native to Mexico

This species has spirals of tubular-shaped flowers which have an attractive lip and droop from the stem. They are brightly coloured and shades available include white, yellow, pink, and crimson. The plant grows to a height of 18 inches (46cm). It is grown in the greenhouse, but when it is in flower it may be brought into the house.

Smithiantha hybrids 'Suttons New Hybrids'.

greenhouse as it must have both warmth and moisture.

Sow the seeds January–March at a temperature of 68–78°F (20–25°C). Just press the seeds into the surface of the seed compost. Germination takes 14–21 days. When the second true leaves have developed, prick out the seedlings singly into 3-inch (8-cm) pots of potting compost. Pot on as necessary.

The flowers bloom May–August.

Sow the seeds January–April at a temperature of 68–78°F (20–25°C). Just press the seeds into the surface of the seed compost. Germination takes 21–42 days. When the seedlings are large enough to handle, prick them out into boxes and grow them on. When they reach a height of about 2 inches (5cm), prick them out singly into 3-inch (8-cm) pots of potting compost. Pot on as necessary.

The flowers bloom June–October.

Snapdragon see *Antirrhinum majus*

Sneezewort see *Achillea ptarmica*

Snow-in-summer see *Cerastium tomentosum*

Snow on the Mountain see *Euphorbia marginata*

Solanum capsicastrum

WINTER CHERRY

Greenhouse shrub
Native to Brazil

This half-hardy evergreen has insignificant white flowers which are followed by dark-green marble-sized fruits which gradually turn to orange-red. It grows to a height of 12–18 inches (30–46cm). It is grown as a pot plant in the greenhouse or as a houseplant. It is especially popular for Christmas decorations as its fruits turn to orange-red at that season.

Sow the seeds February–March at a temperature of 68–78°F (20–25°C) and at a depth of ¼ inch (6mm). Germination takes 14–21 days. When the seedlings are large enough to handle, prick them out into boxes and grow them on. When they are well developed, pot each one into a 3-inch (8-cm) pot of potting compost. Pot on as necessary.

The flowers bloom June–July.

Sorrel, French see *Rumex scutatus*

Spartium junceum

SPANISH BROOM

Hardy deciduous shrub
Native to the Mediterranean region

This shrub has mid-green leaves and long slender branches of golden-yellow flowers. It grows to a height of 8–10 feet (2.4–3 metres). It requires a sunny position and well-drained soil.

Before sowing the seeds, soak them in water for 24 hours. Sow 3 seeds at a depth of ½ inch (1cm) in each 3-inch (8-cm) pot of seed compost. Germination takes up to 50 days. When the seedlings are large enough to handle, reduce them to 1 per pot. Transfer the plants to their permanent positions during the late summer, spacing them far enough apart to allow for their spread of up to 8 feet (2.4 metres).

The flowers bloom July–August.

Above: Solanum capsicastrum.

Left: Solanum capsicastrum 'Red Giant'.

Speedwell see *Veronica teucrium*

Spider Flower see *Cleome spinosa*

Spurge see *Euphorbia marginata*

Stardust see *Gilea lutea*

Star of the Veldt see *Dimorphotheca aurantiaca*

Statice latifolia see *Limonium latifolium*

Statice sinuata see *Limonium sinuatum*

Stock see *Matthiola incana*

Stock, Brompton see *Matthiola incana*

Stock, East Lothian see *Matthiola incana*

Stock, Night-scented see *Matthiola bicornis*

Stock, Perpetual Flowering see *Matthiola incana*

Stock, Ten Week see *Matthiola incana*

Stock, Trysomic see *Matthiola incana*

Stock, Virginian see *Malcolmia maritima*

Straw Flower see *Helichrysum*

Strelitzia reginae
BIRD OF PARADISE FLOWER

Greenhouse perennial
Native to South Africa

This is a striking plant with long-stemmed crested flowers of orange and blue which resemble the head of a bird. It grows to a height of 36–60 inches (91–152cm). It is grown in the greenhouse, either in pots or in the greenhouse border.

Before the seeds are sown soak them in water for 48 hours. Sow the seeds at any time at a temperature of 68–78°F (20–25°C) and at a depth of $\frac{1}{2}$ inch

The dramatic *Strelitiza reginae.*

(1cm). Germination is sporadic, taking from 90 days onwards. When the seedlings are large enough to handle, prick them out singly into 3-inch (8-cm) pots of potting compost. Pot on as necessary. Plants grown from seed will flower 3–4 years after sowing.

The flowers bloom April–May.

Streptocarpus
CAPE PRIMROSE

Greenhouse perennial
Native to South Africa

This plant has trumpet-shaped flowers which are borne well above the foliage. Colours available range from white, through rose and red, to blue and purple. The plant grows to a height of 9–12 inches (23–30cm). It may be grown either as a pot plant in the greenhouse or in the greenhouse border.

For autumn-flowering plants, sow the seeds January–March, at a temperature of 68–78°F (20–25°C) and at a depth of $\frac{1}{8}$ inch (3mm). Germination takes 10–21 days. When the seedlings are large enough to handle, prick them out into boxes of

The Cape primrose, *Streptocarpus.*

Tagetes

T. erecta
AFRICAN MARIGOLD

Half-hardy annual
Native to Mexico

This plant has dark-green foliage and yellow daisylike flowers. There are, however, numerous cultivated varieties, which include single- and double-flowered forms, with colours ranging from yellow, through amber, to orange. They grow to a height of 24–36 inches (61–91cm). There are also semi-dwarf varieties which grow to a height of 12–15 inches (30–38cm). African Marigolds succeed in most soils but require an open sunny position.

Sow the seeds February–March in trays or pots of seed compost at a temperature of 60–68°F (15–20°C) and at a depth of $\frac{1}{4}$ inch (6mm). Germination takes 7–10 days. When the seedlings are large enough to handle, prick them out into boxes and harden them off. Plant them out in the flowering site during May, spacing them 12 inches (30cm) apart. Alternatively, sow the seeds April–May in the flowering site at a depth of $\frac{1}{2}$ inch (1cm). When the seedlings are large enough to handle,

potting compost and grow them on. When they are well developed, plant each seedling in a 3-inch (8-cm) pot of potting compost. Pot on as necessary. For spring-flowering plants, sow the seeds March–April. These plants will flower for the first time the following spring.

The flowers bloom May–October.

Sunflower see *Helianthus annua*

Sweet Cicely see *Myrrhis odorata*

Sweet Corn see *Zea mays*

Sweet Pea see *Lathyrus odorata*

Sweet Pea, Everlasting see *Lathyrus latifolius*

Sweet Rocket see *Hesperis matronalis*

Sweet Sultan see *Centaurea moschata*

Sweet William see *Dianthus barbatus*

There are numerous varieties of marigold available from seedsmen. This is an African marigold (*Tagetes erecta*) from the Jubilee Strain.

Two more examples from the Jubilee Strain:
'First Lady' (*above*) and 'Yellow Galore' (*left*).

thin them out to 12 inches (30cm) apart.

The flowers bloom from July until the first autumn frosts.

T. patula
FRENCH MARIGOLD

Half-hardy annual
Native to Mexico

The foliage is dark green and the flowers are yellow or mahogany-red. The plant grows to a height of about 12 inches (30cm). There are many cultivated varieties which include single, double, and dwarf forms. The dwarf forms grow to a height of 6–9 inches (15–23cm). French Marigolds succeed in most soils but require an open sunny position.

The sowing instructions are the same as those for *T. erecta*.

The flowers bloom from June until the first autumn frosts.

AFRO-FRENCH MARIGOLD

A hybrid between *T. erecta* and *T. patula* is known as the Afro-French Marigold. The colours are similar to those of *T. erecta* and *T. patula* and the plant grows to a height of 10 inches (25cm).

The sowing instructions are the same as those for *T. erecta*.

The flowers bloom from July until the first autumn frosts.

T. tenuifolia pumila
Tagetes signata
MARIGOLD

Half-hardy annual
Native to Mexico

T. t. pumila is a cultivated dwarf form of *Tagetes* and forms a bushy plant

Above: 'Susanna' is a dwarf single-flowered French marigold (*Tagetes patula*).

Above left: 'Boy-O-Boy Mixed' is a dwarf double-flowered French marigold.

Left: 'Goldfinch' is a crested French marigold.

Below left: 'Pascal' is a dwarf single-flowered French marigold.

Below: 'Sunrise' is an F$_1$ hybrid Afro-French marigold.

with neat domes studded with small daisylike flowers in shades of orange or yellow. It grows to a height of 5–7 inches (13–18cm). It thrives on any good garden soil and requires a sunny position.

Sow the seeds February–March at a temperature of 60–68°F (15–20°C). The seeds should be sown at a depth of $\frac{1}{4}$ inch (6mm) in boxes or pots containing seed compost. Germination takes 7–14 days. When the seedlings are large enough to handle, prick them out into trays and grow them on. Harden them off for 2–3 weeks before planting them out in their flowering positions at the end of May. Alternatively, sow the seeds during April in the flowering site at a depth of $\frac{1}{2}$ inch (1cm). When the seedlings are large enough to handle, thin them out to 6–10 inches (15–25cm) apart.

The flowers bloom July–September.

Above: Tagetes tenufolia pumila 'Golden Gem'; *right: T. t. pumila* 'Tangerine Gem'. Both varieties make good edging plants and can be grown on any good garden soil.

Tanacetum vulgare

TANSY

Hardy perennial
Native to Europe

The leaves are dark green in colour, and yellow buttonlike flowers appear July–August. The plant grows to a height of 36–48 inches (91–122cm). It grows best on a moist, loamy soil. As Tansy has a creeping rootstock, confine it to a secluded part of the garden or plant it in a sunken metal bucket from which the bottom has been removed.

Sow the seeds in open ground in early spring or during the autumn. Sow them in shallow drills about $\frac{1}{4}$ inch (6mm) deep. When the seedlings are large enough to handle, thin them out to 12–24 inches (30–61cm) apart.

Tansy leaves may be shredded and used in omlettes or in sandwiches with cream cheese. The flowers may be dried for use in winter flower arrangements.

Tassel Flower see *Emilia flammea*

Temple Bells see *Smithiantha hybrids*

Among the plants in this border edging a pathway are thyme, salvia and bells of Ireland (*Molucella laevis*).

Thistle, Globe see *Echinops ritro*

Thistle, Ornamental see
Onopordum acanthium

Thrift see *Armeria maritima*

Thunbergia alata

BLACK-EYED SUSAN

Half-hardy annual
Native to South Africa

This climbing plant has flowers which
are white, yellow, or orange with a
black eye. It grows to a height of 48
inches (122cm). It is grown in a cool
greenhouse or a sunny position in the
garden. It is suitable also as a trailing
plant in hanging-baskets. It grows
very quickly and is in flower over a
long period.

Sow the seeds March–April at a
temperature of 68–78°F (20–25°C).
The seeds should be sown in 3-inch
(8-cm) pots of seed compost at a depth
of $\frac{1}{4}$ inch (6mm), allowing 3–4 seeds
per pot. Germination takes 14–28
days. Pot on as necessary.

The flowers bloom June–
September.

Thunbergia alata, a very effective climbing
plant. The variety shown below is 'Susie'.

Thymus

T. serpyllum
SHEPHERD'S THYME, WILD THYME

Hardy perennial
Native to Europe

The grey-green foliage is covered with heather-purple flowers. It grows to a height of 3 inches (8cm). It is especially suitable for ground cover and may be planted between paving stones as it emits a pleasant odour when crushed. It requires a sunny position and thrives on any well-drained soil.

Sow the seeds April–June in a cold frame or cool greenhouse at a depth of $\frac{1}{8}$ inch (3mm). Germination takes 14–28 days. When the seedlings are large enough to handle, prick them out into boxes and grow them on. Transfer the plants to their permanent positions during September, spacing them far enough apart to allow for their spread of up to 24 inches (61cm).

The flowers bloom June–August.

T. vulgaris
COMMON THYME, GARDEN THYME

Hardy perennial
Native to southern Europe

A miniature evergreen bush with tiny dark-green leaves and clusters of tube-shaped mauve flowers. It grows to a height of 12 inches (30cm). It requires a sunny position and thrives on any well-drained garden soil.

Sow the seeds April–May in the open ground at a depth of $\frac{1}{8}$ inch (3mm). When the seedlings are large enough to handle, thin them out to 12 inches (30cm) apart.

The leaves are always used in *bouquet garni* and may be used in recipes that include wine, shell-fish or pork.

Tickseed see *Coreopsis*

Toadflax see *Linaria maroccana*

Tobacco Plant see *Nicotiana*

Torenia fournieri
WISHBONE FLOWER

Greenhouse annual
Native to Vietnam

This plant has pale-green leaves and violet-blue flowers which resemble Snapdragons. It grows to a height of 12 inches (30cm). It is usually grown as a decorative pot plant in the greenhouse.

Sow the seeds February–April at a temperature of 60–68°F (15–20°C) and at a depth of $\frac{1}{8}$ inch (3mm). Germination takes 14–21 days. When the seedlings are large enough to handle, prick them out singly into 3-inch (8-cm) pots of potting compost. Pot on as necessary until the plants are in 5-inch (13-cm) pots.

The flowers bloom July–September.

Touch-me-not see *Impatiens balsamina*

Trachelium caeruleum

Hardy biennial, best treated as a half-hardy annual
Native to Europe

The large clustered flowerheads resemble mauve Gypsophila. The plant grows to a height of 24 inches (61cm). It may be grown in a pot in a cool greenhouse or planted outside in a warm, sunny position.

Sow the seeds January–March at a temperature of 60–68°F (15–20°C) and at a depth of $\frac{1}{8}$ inch (3mm). Germination takes 14–21 days. When the seedlings are large enough to handle, prick them out into boxes. Harden off those that are to be planted outside and transfer them to their flowering positions during May, spacing them 12 inches (30cm) apart. Prick out singly those seedlings that are to be grown on as pot plants into 3-inch (8-cm) pots of potting compost. Pot on as necessary.

For flowers the following spring, sow the seeds June–July at a depth of ¼ inch (6mm) in a nursery bed. When the seedlings are large enough to handle, thin them out to 6 inches (15cm) apart and grow them on until the autumn. Transfer the plants to their flowering positions September–October, spacing them 12 inches (30cm) apart.

The flowers bloom May–September.

Tritoma see *Kniphofia*

Tropaeolum
Native to Peru

T. majus
NASTURTIUM

Hardy annual

This plant has circular mid-green leaves and yellow or orange flowers. It grows to a height of 8 feet (2.4 metres). Cultivated varieties have flowers in many colours including red, pink, and maroon as well as yellow and orange. Dwarf varieties, which grow to a height of 10–18 inches (25–46cm), are also available. Nasturtiums climb up and through hedges or trail over banks. The dwarf varieties are ideal for window-boxes and hanging-baskets. Nasturtiums thrive on poor, dry soil.

Sow the seeds April–May in the flowering site at a depth of ½ inch (1cm). Germination takes 10–14 days. When the seedlings are large enough to handle, thin them out to 15 inches (38cm) apart. Thin out dwarf varieties to 12 inches (30cm) apart.

The flowers bloom June–September.

Nasturtium leaves and flowers may be used in salads.

Tropaelum majus: (*above*) the semi-tall 'Double Gleam' variety and (*below*) the dwarf semi-double-flowered variety 'Red Roulette'.

Jewel Mixed', another dwarf semi-double-flowered variety of *Tropaelum majus*.

T. peregrinum
T. canariense
CANARY CREEPER

Half-hardy perennial usually grown as
a hardy annual

This rapid-growing creeper has
canary-yellow flowers. It may grow to
a height of 12 feet (3.6 metres) in one
season. It succeeds in either a sunny or
shady position but prefers a rich soil.
It may be grown as a trailing plant in
hanging-baskets or window-boxes.
    Sow the seeds February–March at a
temperature of 60–68°F (15–20°C) and
at a depth of $\frac{1}{2}$ inch (1cm).
Germination takes 21–35 days. When
the seedlings are large enough to
handle, prick them out into boxes and
harden them off. Transfer the plants to
their flowering positions during May,
spacing them 36 inches (91cm) apart.
Alternatively, sow the seeds
April–May at a depth of $\frac{1}{2}$ inch (1cm)
in the flowering site. When the
seedlings are large enough to handle,
thin them out to 36 inches (91cm)
apart.
    The flowers bloom July–October.

Umbrella Plant see Cyperus alternifolius

Ursinia anethoides is a very free-flowering
species.

harden them off. Transfer the plants to
their flowering positions during May,
spacing them 12 inches (30cm) apart.
Alternatively, sow the seeds
April–May in the flowering site at a
depth of $\frac{1}{4}$ inch (6mm). When the
seedlings are large enough to handle,
thin them out to 12 inches (30cm)
apart.
    The flowers bloom June–September.

Vaccaria pyramidata see Saponaria vaccaria

Valerian see Centranthus ruber

Ursinia anethoides
Half-hardy perennial usually treated
as a half-hardy annual
Native to South Africa

This bushy plant has light-green
foliage and orange daisylike flowers
with contrasting centres. It grows to a
height of 9 inches (23cm). It requires a
sunny position and well-drained soil.
It may be grown as a pot plant.
    Sow the seeds February–April at a
temperature of 60–68°F (15–20°C) and
at a depth of $\frac{1}{8}$ inch (3mm).
Germination takes 10–14 days. When
the seedlings are large enough to
handle, prick them out into boxes and

Verbascum phoeniceum
MULLEIN

Hardy perennial
Native to the Levant

This cottage-garden plant has graceful
spikes of white, pink, purple, mauve,
or blue flowers. It grows to a height of
24 inches (61cm). It requires a sunny
position and well-drained soil and may
be planted in borders or among
shrubs.
    Sow the seeds April–July at a depth
of $\frac{1}{4}$ inch (6mm) in a seed bed.
Germination takes 14–28 days. When
the seedlings are large enough to
handle, thin them out to 6 inches

(15cm) apart and grow them on until the autumn. Transfer the plants to their permanent positions during September, spacing them 15–18 inches (38–46cm) apart.

The flowers bloom May–September.

Verbena
VERVAIN

V. × hybrida
V. hortensis

Half-hardy perennial usually treated as a half-hardy annual
Derived from species native to South America

This hybrid species produces bushy plants with mid-green foliage. The flowers are bright and colours include scarlet, blue, carmine, and white. The plant grows to a height of 12–15 inches (30–38cm). Dwarf varieties are also available and grow to a height of 9–12 inches (23–30cm). Verbena requires a well-cultivated soil and an open sunny position. It is particularly suitable for use as a bedding or edging plant, and

Verbena × hybrida, the garden varieties of vervain are brilliantly coloured and their flowers last until the first frost of autumn.

Verbena × hybrida 'Springtime' is a dwarf spreading variety.

may be used also in window-boxes.

Sow the seeds January–March at a temperature of 60–68°F (15–20°C). The seeds should be sown in pots or boxes of seed compost at a depth of $\frac{1}{8}$ inch (3mm). Do not overwater the compost as Verbena germinates best when the compost is kept just moist. Germination takes 10–21 days. When the seedlings are large enough to handle, prick them out into boxes and grow them on. Harden them off and plant them out during May, spacing them 12 inches (30cm) apart.

The flowers bloom from June until the first autumn frosts.

V. rigida
V. venosa

Hardy perennial
Native to South America

This species has dark-green foliage and mauve flowers. It grows to a height of 12–24 inches (30–61cm). 'Alba', an off-white variety, is also available. It requires an open sunny position and well-cultivated soil.

The sowing instructions are the same as those for *V. × hybrida*, but when the plants are transferred to their

Above: Verbena venosa.

Below: Vinca rosea 'Little Gem Mixed'.

flowering positions, space them 15 inches (38cm) apart.

The flowers bloom July–October.

Veronica teucrium
SPEEDWELL

Hardy perennial
Native to Europe

This plant forms loose hummocks of erect stems and bears spikes of bright-blue flowers. It grows to a height of 12 inches (30cm). It requires a sunny position and well-drained soil.

Sow the seeds May–July at a depth

of $\frac{1}{4}$ inch (6mm) in a seed bed. Germination takes 14–28 days. When the seedlings are large enough to handle, thin them out to 9 inches (23cm) apart and grow them on until the autumn. Transfer the plants to their permanent positions during September, spacing them far enough apart to allow for their spread of up to 24 inches (61cm).

The flowers bloom June–August.

Vervain see *Verbena*

Vinca rosea
Catharanthus roseus
MADAGASCAR PERIWINKLE

Greenhouse perennial
Native to Malagasy

This plant has dark-green foliage, and pink or white flowers. It grows to a height of 12 inches (30cm). Although primarily a greenhouse plant, it may be grown out of doors in warm sheltered areas.

Sow the seeds during March at a temperature of 60–64°F (15–18°C) and at a depth of $\frac{1}{8}$ inch (3mm). When the seedlings are large enough to handle, prick them out singly into 3-inch (8-cm) pots of potting compost. Pot on as necessary. If the plants are to be transferred to a position in the garden, prick out the seedlings into boxes and harden them off in a cold frame. When all danger of frost has passed (i.e., in late May or early June), transfer the plants to their flowering positions, spacing them 12 inches (30cm) apart.

The flowers bloom April–October.

Viola

V. odorata
SWEET VIOLET

Hardy perennial
Native to Europe

This species of Violet has mid-green foliage and large fragrant flowers in shades of purple or white. It grows to a height of 4–6 inches (10–15cm). It requires a moist but well-drained soil and succeeds in either a sunny or partially shaded position. It spreads by means of runners and is suitable for bedding schemes or for use as an edging to paths.

Sow the seeds September–November at a depth of $\frac{1}{4}$ inch (6mm), and over-winter them in a cold frame or cool greenhouse. Germination will take place the following spring. When the seedlings are large enough to

Above: Viola x *Williamsii* 'Campanula Blue'; *below: Viola* x *Williamsii* 'Large-flowered Mixed'. These plants are similar to pansies, but are smaller.

handle, prick them out into boxes and grow them on until the autumn. Transfer the plants to their permanent positions during September, spacing them far enough apart to allow for their spread of up to 12 inches (30cm).

The flowers bloom February–April.

V. × *williamsii*
VIOLA, TUFTED PANSY

Hardy perennial

The flowers of this species resemble Pansies but are smaller. Cultivated varieties are available in a wide range of colours including blue, violet, yellow, and white. The plant grows to a height of 4–6 inches (10–15cm). Violas grow best in a cool moist position and prefer a well-cultivated soil.

Sow the seeds June–July at a depth of $\frac{1}{4}$ inch (6mm) in a seed bed. Germination takes 10–21 days. When the seedlings are large enough to handle, thin them out to 12 inches (30cm) apart. Transfer the plants to the flowering site during September, spacing them 12 inches (30cm) apart. Alternatively, for flowers the same year, sow the seeds January–March at a temperature of 50–60°F (10–15°C) and at a depth of $\frac{1}{4}$ inch (6mm). Germination takes 10–21 days. When the seedlings are large enough to handle, prick them out into boxes and harden them off. Plant them out during May, spacing them 12 inches (30cm) apart.

The flowers bloom May–September.

V. × *wittrockiana*
PANSY

Hardy biennial

This garden hybrid has been bred from *V. tricolor* (Heart's Ease). The brilliantly coloured flowers are 1–4 inches (2.5–10cm) in width, depending on the variety. Colours include yellow,

red, blue, white, and violet. The plant grows to a height of 6–9 inches (15–23cm). Pansies thrive on cool, moist, well-cultivated soils and tolerate both full sun and partial shade.

Sow the seeds June–July at a depth of $\frac{1}{4}$ inch (6mm) in a seed bed. When the seedlings are large enough to handle, thin them out to 4 inches (10cm) apart. Transfer them to the flowering site September–October,

Above and left: examples of the numerous varieties of *Viola* x *Wittrockiana* (pansy) that are available from seedsmen.
Above: 'Super Chalon Giants Mixed', which have unusual ruffled flowers.
Left above: 'Crystal Bowl Mixed', an F₁ hybrid.
Far left below: 'Imperial Orange Prince', an F₁ hybrid.
Left below: 'Colour Festival Mixture', an F₂ hybrid.

spacing them 9–12 inches (23–30cm) apart. They will flower the following year.

The flowers bloom May–July.

Violet, African see *Saintpaulia ionantha*

Violet, Sweet see *Viola odorata*

Viper's Bugloss see *Echium lycopsis*

Virginian Stock see *Malcolmia maritima*

Virginian Creeper see *Parthenocissus tricuspidata 'Veitchii'*

Viscaria vulgaris see *Lychnis viscaria*

Wallflower see *Cheiranthus cheiri*

Wallflower, Alpine see *Erysimum alpinum*

Wallflower, Fairy see *Erysimum alpinum*

Wallflower, Siberian see *Cheiranthus × allionii*

Wishbone Flower see *Torenia fournieri*

Woodruff, Annual see *Asperula orientalis*

Xeranthemum annum

Hardy annual
Native to southern Europe

The daisylike flowers are available in both single and double forms and in many colours including white, rose, lilac, and purple. It grows to a height of 24 inches (61cm). It requires a well-drained soil and sunny position. This 'everlasting' plant is greatly prized by flower arrangers for use in winter decorations.

Sow the seeds March–April at a depth of $\frac{1}{4}$ inch (6mm) in the flowering site. Germination takes 10–14 days. When the seedlings are large enough to handle, thin them out to 12 inches (30cm) apart.

The flowers bloom June–July.

Yarrow see *Achillea*

Xeranthemum annum, an 'everlasting' flower

Zea mays
SWEET CORN

Half-hardy annual
Native to the USA

Ornamental Sweet Corn is grown for its decorative multicoloured seeded

cobs which, when ripe, range in colour
from golden-yellow to purple. Foliage
of some varieties is decorative enough
to be used in summer bedding
schemes. Both seed cobs and foliage
are useful to flower arrangers as they
may be dried for use in winter
decorations. The plant grows to a
height of 24–48 inches (61–122cm),
depending on the variety. It requires
an open sunny position and well-
cultivated soil, preferably one that has
been enriched with organic material
during the previous year.

Sow the seeds March–April at a
temperature of 60–68°F (15–20°C) and
at a depth of ½ inch (1cm).
Germination takes 7–10 days. When
the seedlings are large enough to
handle, prick them out into boxes and
harden them off. Transfer the plants to
their flowering positions during May,
spacing them 18 inches (46cm) apart.
Alternatively, sow the seeds in the
flowering site at a depth of 1 inch
(2.5cm) during May. Germination
takes 7–14 days. When the seedlings
are large enough to handle, thin them
out to 18 inches (46cm) apart.

Above: The multi-coloured cobs of *Zea mays* are
very decorative. *Below:* A double-flowered
variety of *Zinnia elegans*

Zinnia elegans

Half-hardy annual
Native to Mexico

This is the most popular species of
Zinnia. It has mid-green foliage and
brightly coloured, daisylike flowers.
Among the colours available are white,
purple, orange, yellow, red, and pink.
The variety 'Envy' has lime-green
flowers. Zinnias grow to a height of
24–30 inches (61–76cm). Semi-double
and double-flowered varieties are
available, as are dwarf strains which
reach a height of 6–15 inches
(15–38cm), depending on the variety.
Zinnias must be planted where they
will be in full sunshine and grow best
on a good well-drained soil.

Sow the seeds March–April at a
temperature of 60–68°F (15–20°C) and
at a depth of ¼ inch (6mm).
Germination takes 7–14 days. When
the seedlings are large enough to
handle, prick them out into boxes, and
grow them on. Harden them off for
2–3 weeks. Transfer the plants to their
flowering positions early in June,
spacing them 12 inches (30cm) apart.
Alternatively, sow the seeds in the
flowering site at a depth of ½ inch
(1cm) during May. Germination takes
10–14 days. When the seedlings are
large enough to handle, thin them out
to 12 inches (30cm) apart.

The flowers bloom July–September.

Above: Zinnia elegans 'Pulchino', a dwarf variety.
Below: 'Pink Ruffles', an F₁ hybrid

Above: The green zinnia 'Envy'.
Below: A single-flowered zinnia, 'Chippendale Daisy'.

Below: 'Yellow Ruffles', an F₁ hybrid.

Below: 'Peter Pan Mixed', an F₁ hybrid.

HERB GARDENS

Herb gardens are beautiful as well as useful. They are traditionally
laid out in small plots, often divided by paths or paving stones,
with the different plants easily accessible for picking.

Below: The herb garden at Sissinghurst
Castle, Kent.

Below: An example of groups of herbs divided
by paving stones.

Left: The herb garden at Hyde Hall, Essex.
Below: The medicinal herb garden at Acorn Bank, Cumbria.

Below: The herb garden at Rosemoor, Devon, with paving and gravel paths.

Below: The herb garden at Scotney Castle. Kent.

GLOSSARY

Annual

A plant which germinates, flowers, produces seeds and dies within a year. Examples are *Alyssum maritimum, Centaurea cyanus* (Cornflower) and *Malcolmia maritima* (Virginian Stock).

Bedding Out

The bedding plants should be gently eased out of their box. This is easily done if you bang the sides of the box first. Once the plants are out of the box separate them carefully so that each one has a good root. Use a trowel to dig a hole which is big enough for the plant's root ball, insert the plant then replace the soil slightly above the level of the original soil. Firm the soil down well and water it. (*See also* Bedding Plants.)

Bedding Plants

A hardy or half-hardy annual, biennial or perennial which is used for display during the summer. These plants can usually be bought ready for bedding out from a nursery man. (*See also* Bedding Out.)

Biennial

A plant which has a life cycle spread over two years. During the first year the plant produces leafy growth; during the second year it produces flowers and seeds and then dies. Examples are *Dianthus barbatus* (Sweet William), *Lunaria annua* (Honesty) and *Oenothera trichocalyx* (Evening Primrose). Some perennials (*qv*) are treated as biennials, for example *Cheiranthus cheiri* (Wallflower).

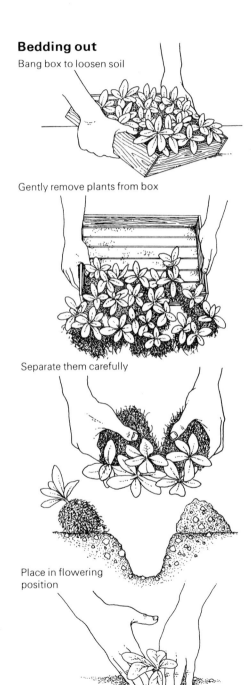

Bedding out
Bang box to loosen soil

Gently remove plants from box

Separate them carefully

Place in flowering position

Capsule

A dry fruit which usually contains loose seeds.

Cloche

Cloches are available in a number of shapes and sizes and are used for protecting early plants. They were traditionally made of glass but plastic ones are now becoming popular as they are cheaper and lighter than glass. Tunnel cloches are made from corrugated PVC held in position by wire hoops or from lightweight polythene which is laid over a series of hoops and held in place by a second set of hoops. Cloches made of corrugated PVC are probably the most practical as they last longer than glass, which breaks easily, or lightweight polythene, which deteriorates in sunlight.

The cloche should be put in position about two weeks before it is needed so that the soil under it warms up. Should there be any danger of frost the cloche should be covered with several thicknesses of newspapers which can be held in position with stones.

Clone

The descendants of a plant that has only been propagated asexually. Many clones are sterile, but even if the plants set seed the offspring would not be regarded as part of the clone.

Compost

This word is used in two ways: a) to refer to an organic fertilizer made by composting vegetable waste; b) to refer to the soil mixtures in which seeds are sown or pot plants are grown (*see* Growing Medium).

One can make compost simply by piling up vegetable waste in a corner of the garden. However, as such heaps are inclined to spread and look untidy,

Cloches

Tent cloche

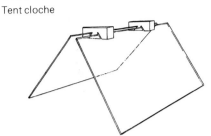

Barn cloche

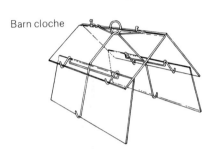

Moulded plastic cloche

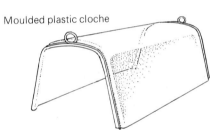

Corrugated plastic tunnel cloche

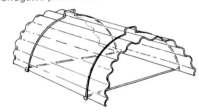

Soft plastic tunnel cloche

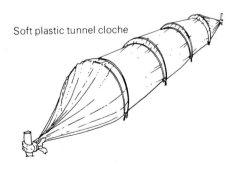

Types of compost bin

Wire netting

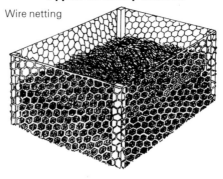

Wood

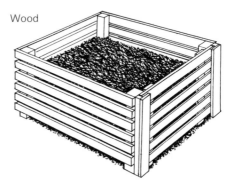

Commercially available bin with sliding panels

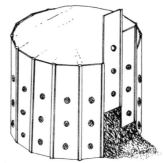

Plastic bag

a. Insert wire hoops in ground
b. Attach one end of plastic to a peg; pull cover
    over hoops
c. Attach other end of plastic to a peg; clip the
    outer hoops on to the inner hoops

it is advisable to contain the compost heap in some way. Suitable containers can be made out of chicken wire or wood or a special container can be bought.

Almost anything can be composted including leaves, grass clippings, egg shells, even shredded newspaper. Annual weeds can be composted but not perennial weeds such as Dandelion and Bindweed, which should be burned. Woody materials, e.g. tree prunings, should not be used as they will not decompose. *Never* put cooked kitchen scraps on the compost heap — they will encourage rats and other vermin. Do not build the heap using only one form of waste, e.g. grass clippings, as it will become sour.

Sprinkle some proprietary compost activator over each 9–12 inch (23–30cm) layer of waste. Turn the compost occasionally so that the materials on the outside of the heap are moved to the middle. Keep the compost covered, by using a polythene sheet held down by stones. Compost bins bought ready-made usually have lids and it is not necessary to turn the compost in such bins.

The compost is ready for use when it is dark brown and crumbly. In summer it should be ready in about 2–4 months; in winter it will be ready in about 4–8 months.

If you are not a do-it-yourself enthusiast and do not want to invest in a ready-made bin you will find that the black plastic bags used for rubbish can be used to make compost. Simply follow the instructions given above but ensure before you begin that you make holes in the bottom of the bag for drainage and holes in the sides to let air in.

Containers see Pots

Corolla

The ring of petals in the flower as a whole.

Corona

The trumpet- or cup-shaped part of the flower which lies between the petals and the stamens.

Cotyledon

The first seedling leaf or leaves to appear at germination. These leaves are frequently different in shape from the plants' adult leaves. (*See also* Dicotyledon; Monocotyledon.)

Crocks

These are broken pieces of a clay pot which are placed over the drainage hole of a container.

Cultivar

A variant of a plant, either species or hybrid, which has arisen in cultivation and is not known in the wild. Wild variants are known either as varietas, abbreviated var. or forma, abbreviated f. The latter are given Latin names, while cultivars (abbreviated cv.) are given names in the language of the country in which they were raised.

Deciduous

A tree or shrub that loses its leaves at the end of each growing season.

Dicotyledon

Plants in this group have two seed leaves (*see* Cotyledon).

Digging

Digging is essential to the creation of a fertile soil. It should be carried out in autumn so that the soil can be broken down by frost and rain during the winter. The process aerates the soil and manure or compost can be incorporated at the same time.

Single digging

a. Take out a trench at one end of the plot;
   move soil to the other end of the plot
b. Fill in the trench with soil from the next
   trench. Repeat the process until you reach
   the end of the plot
c. Manure or compost can be added as you dig
   each trench – place it at the bottom of the
   trench
d. Fill in the final trench with soil taken from
   the first trench

179

a

b   c

d

Double digging

a. Dig first trench one spit deep and about 18
   inches (45 cm) wide. Use a fork to break up
   the base of the trench to the depth of
   another spit
b. Add manure by placing it over the loosened
   base of the trench
c. Fill in the first trench with soil from
   succeeding trenches until you reach the end
   of the plot
d. Fill in the final trench with soil taken from
   the first trench

There are two types of digging: single digging which consists of turning over the top spit of soil and double digging which consists of turning over the soil to a depth of two spits.

If you have never done any digging before the wisest course is to do no more than an hour's work at a time. Ensure that when you insert the spade in the soil it is vertical as digging with the spade at an angle results in the soil being only shallowly cultivated.

Disk (Disc) Floret see Floret

Double Flowers

Abnormal flowers in which the stamens and/or pistils have been transformed into petals. Fully double flowers are, therefore, completely sterile, but some double flowers still have a few stamens and often undamaged pistils, from which seed can be raised. In semi-double flowers only a few of the sexual organs have become petaloid. Plants of the Daisy family with all ray florets are often termed double, but this is inaccurate; there is no such thing as a single dandelion, for example.

Drawn

A plant which is growing in a group that is too closely packed or in a poorly lighted position is inclined to become long and thin and is often pallid in colour. Such a plant is said to be 'drawn'.

Drill

A straight, shallow u- or v- shaped furrow in which seeds are sown. The easiest way to make a drill is with a rake or hoe.

Evergreen

A tree or shrub which bears foliage throughout the year.

F_1

A hybrid plant which is the first generation of a controlled crossing of parent plants. Seeds from F_1 hybrids do not come true and so the crossing has to be repeated to reproduce the original hybrid.

F_2

A hybrid plant which is the second generation of a controlled crossing of parent plants.

Fertilizers

Fertilizers supply nutrients which may not be present in sufficient quantities in the soil. The nutrients are essential to the plant's growth. There are two types of fertilizers: organic and

Making a drill

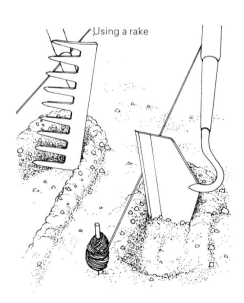

Using a rake

Using a hoe

inorganic. Organic fertilizers (e.g. bonemeal, dried blood) are discussed in the section on Manure and so here we will discuss inorganic fertilizers.

A general fertilizer, i.e. one containing a balanced blend of phosphorus, nitrogen and potassium, is the easiest type to use. Applying individual fertilizers such as sulphate of ammonia or superphosphate of lime involves careful measurement of the amount used and can lead to an imbalance of nutrients in the soil.

The fertilizer should not be applied near the plant's stem or leaves unless you are using a foliar feed which must be applied to the leaves. The best position for the fertilizer is in a circle around the plant in line with the outer edge of the plant's upper growth.

A general fertilizer should not be dug into the soil. Apply it before the seeds are sown or before planting out and then use it occasionally as the plant is growing. Do not use a fertilizer during the winter as the plants are not growing then. Follow the instructions on the packet as to the amount to be applied over a given area. Do not apply a fertilizer when the ground is dry—this also applies to liquid fertilizers.

Liquid fertilizers are more expensive than dry ones but are more easily absorbed by the plants and therefore give quicker results. However, they can be washed out of the soil very quickly when it rains.

Floret

An individual flower which forms part of a large flowerhead or inflorescence. Disk florets are tubular, petalless florets found in many plants of the Daisy family. Sometimes they are surrounded by the more conspicuous ray florets, as in the Common Daisy. Groundsel is composed entirely of Disk florets, while some Chrysanthemums are composed entirely of ray florets.

Flower

The reproductive organ of the plant.

Frame

A frame, usually referred to as a cold frame, is a topless, bottomless box with a cover or covers (light or lights) made of glass or plastic. The frame can be made of wood, brick, concrete blocks or even, in an emergency, turves or bales of straw. It is generally used to harden off plants. (*See* Hardening Off.)

Friable

A soil which is crumbly and therefore can be easily worked.

Fruit

The mature ovary which contains the ripened seeds. The fruits may be dry pods, e.g. the Pea, or capsules, e.g. the Poppy, or soft and fleshy, e.g. the Tomato. (*See* also Capsule.)

Fungicides see Sprays

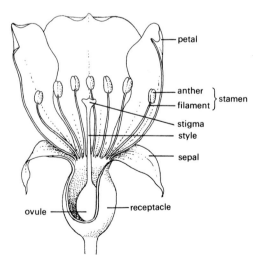

petal

anther
filament } stamen

stigma
style

sepal

ovule

receptacle

Cross-section of a regular flower

Types of cold frame

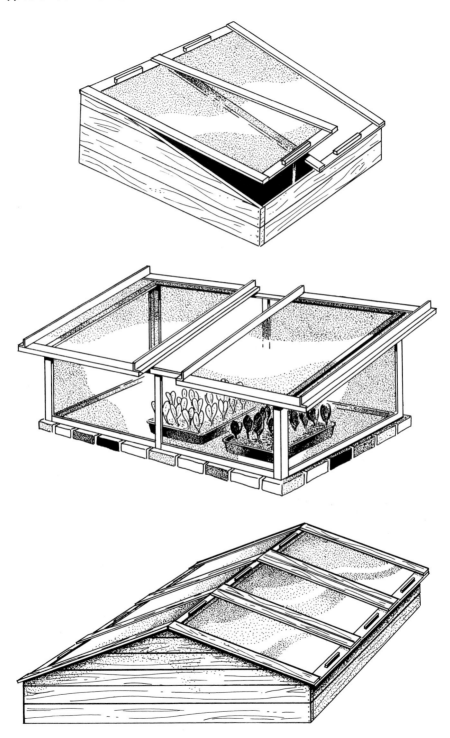

A seedling in the various stages of germination

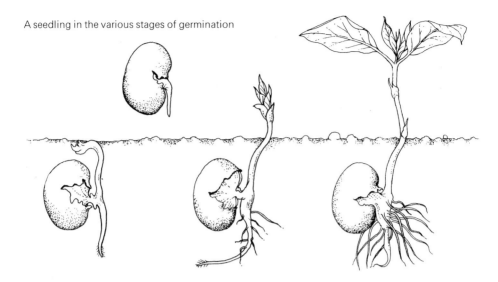

Genus

A division of a plant family which is based on the plant's botanical characteristics. A plant's genus is indicated by its first botanical name. For example, Cornflower belongs to the genus *Centaurea*. The plural form of genus is genera.

Germination

The first stage in the development of a plant from a seed.

Greenhouses

Many different shapes and sizes of greenhouse are available and have cedarwood or aluminium frames. Among the most popular types are the following:

The traditional vertical-sided greenhouses are free-standing and are often timber-clad up to the level of the staging.

Dutch light greenhouses have sloping sides and the large glass panels reach the ground on all sides.

Lean-to greenhouses are three-sided, the fourth side being formed by the building against which they are built. If it is built against a house wall it will be warmer than other types of greenhouse.

Circular greenhouses are becoming increasingly popular. They have the advantage that the gardener can work on all parts of the greenhouse from the centre of the greenhouse simply by turning around which wastes less space than in other houses which need space along their length for the gardener to work in.

An east-west alignment is usually considered to be ideal for the greenhouse as it then gets light throughout the day. Obviously the greenhouse should not be sited under trees or where it will be overshadowed by any other building, nor should it be sited too near a fence or hedge as this may make maintenance difficult.

A greenhouse must have a good ventilation system with two or more roof ventilators and at least one at a lower level. Automatic ventilators are available and are very useful for gardeners who are out all day. They must, however, be connected to the main electricity supply. The other essential is a regular and plentiful supply of water. Again, automatic systems are available. Capillary

Types of greenhouse

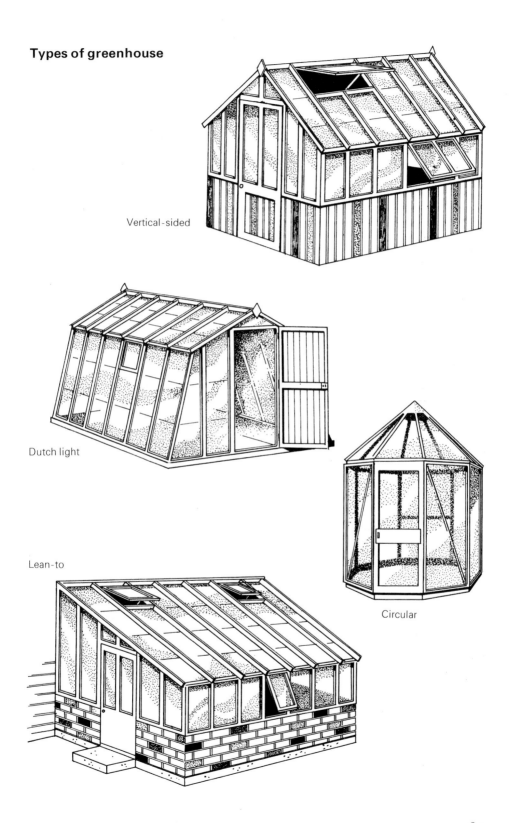

Vertical-sided

Dutch light

Lean-to

Circular

Greenhouse equipment

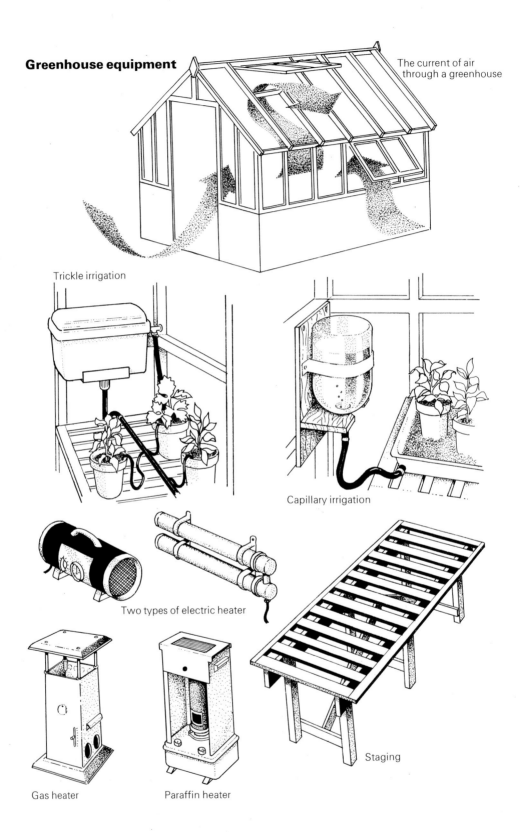

The current of air through a greenhouse

Trickle irrigation

Capillary irrigation

Two types of electric heater

Gas heater

Paraffin heater

Staging

irrigation is a system by which water flows from a tank into a trough from which it is drawn into a sand tray by means of a wick. A second capillary system involves standing the plant pots on an absorbent mat which takes water from a tank. This type of watering is really only suitable for very small plants or for seedlings. Trickle irrigation is suitable for larger plants. A pipe is laid near the plants and nozzles from it are placed in each pot releasing into them a steady stream of water.

The greenhouse will probably need some form of shading to protect the plants from direct sunlight during the hottest periods. Slatted blinds are the most efficient way of supplying shade, but a spray is available which can be applied to the glass. It becomes opaque in bright sunlight but transparent on dull days.

Unless you intend to use the greenhouse as a cold house it will need some form of heating during the winter. Gas, paraffin and electric heaters are all available. Electric heating is the most expensive form, but is automatic and more versatile than the other systems. The conventional form of heating greenhouses is by hot-water pipes from a boiler, which may be fired by gas, oil or solid fuel. This is not suitable for small structures, but where a lean-to greenhouse is placed against the wall of the dwelling it may be possible to extend the central heating system to warm the lean-to.

Greenhouse staging, or slatted benches, can be made by the handyman using $2\frac{1}{2}$ inch (5cm) timber for the legs and 2×1 inch (5×2.5cm) timber for the framework of the bench area. Battens made from $2\frac{1}{2} \times \frac{1}{2}$ inch (5×1cm) timber are nailed to the top to form the slats. The bench is strengthened by the addition of crossbracing, made from 2×1 inch (5×2.5cm) timber, between the legs and top. The staging should be placed on a firm surface, e.g. stone slabs, to ensure that it remains steady.

The greenhouse must be kept clean and should be fumigated with pesticidal and fungicidal smoke 'bombs' at the end of each growing season. The smoke reaches all parts of the greenhouse and is harmless to food crops. The bombs must be used in calm weather, but not in bright sunshine. The ventilators must be shut and any gaps blocked before the bombs are lit. Light the first bomb at the end of the greenhouse furthest from the door and work backwards towards the door. Close and lock the door. If it has no lock, put a notice on it. The following day open the door, then wait for a few minutes before opening the ventilators.

Disinfectant can be used to clean the interior of the greenhouse and don't forget to clean both sides of the glass panes so that as much light as possible can reach the plants. This is best done before spring sowing begins.

Ground Cover

Plants referred to as 'ground cover plants' are used to cover the soil, usually under trees or between shrubs. Besides bringing colour to the planting scheme they provide a dense weedproof cover. *Hypericum calycinum* (St John's Wort) is an excellent ground cover plant.

Growing Medium

Basically two types are available; seed compost, which is used specifically for seed sowing and for rooting cuttings, and potting compost, which is used for growing pot plants. Both types of growing medium are usually based on peat, with sand, soil nutrients and sometimes loam added. They are commercially available, the best

known being the John Innes
Composts, but the keen gardener may
wish to make up his own mixtures.

The formulae are:

John Innes Seed Compost (JIS)
2 parts by volume loam
1 part by volume peat
1 part by volume sand
plus for every 8 gallons (36 l) of the
mixture:
1½oz (43g) single superphosphate
¾oz (22g) ground chalk, ground
limestone or whiting

John Innes Potting Compost (JIP1)
7 parts by volume loam
3 parts by volume peat
3 parts by volume sand
plus for every 8 gallons (36 l) of the
mixture:
4 oz (113g) John Innes Base Fertilizer
¾ oz (22g) ground chalk, ground
limestone or whiting.
(For seedling plants transplanted from
seed compost.)

*John Innes Potting Compost No. 2
(JIP2)*
7 parts by volume loam
3 parts by volume peat
3 parts by volume sand
plus for every 8 gallons (36 l) of the
mixture:
8 oz (226g) John Innes Base Fertilizer
1½ oz (44g) ground chalk, ground
limestone or whiting.
(For plants grown in JIP1 when
transplanted or potted for the second
time.)

*John Innes Potting Compost No. 3
(JIP3)*
7 parts by volume loam
3 parts by volume peat
3 parts by volume sand
plus for every 8 gallons (36 l) of the
mixture:
12 oz (339g) John Innes Base Fertilizer
2¼ oz (66g) ground chalk, ground
limestone or whiting

John Innes Base Fertilizer
2 parts by weight hoof and horn meal
2 parts by weight single phosphate
1 part by weight sulphate of potash

It must be noted that differences in the
loam used in loam-based composts can
lead to variable results. As a result, a
number of soil-less composts have
been developed under such names as
Levington and BIO. They are
basically mixtures of peat and sand
with added fertilizers and lime.

Half-hardy

This refers to plants from tropical and
sub-tropical regions which require
protection during the winter when
grown in temperate climates. It may
also refer to certain shrubs and
herbaceous perennials which will
survive average—but not severe—
winters out of doors if grown in
sheltered positions or in regions which
have a mild climate. (*See also* Hardy.)

Hardening Off

This is the method adopted to
accustom plants to a cooler
environment than the one in which
they germinated. For example, plants
that are germinated in a greenhouse
are hardened off by being placed in a
cold frame during late spring. The
lights of the frame are raised further
each day to allow more air in until they
are completely removed. The plants
can then be transferred to their
permanent position. (*See also* Frame.)

Hardy

This refers to plants which are capable
of surviving, indeed thriving, under
the natural environment given to
them. (*See also* Half-hardy.)

Herbaceous

A plant which does not form a woody stem, remaining soft and green throughout its life. The term is used to refer to annuals, biennials and perennials. Such plants die down to the ground each winter.

Herbicides see Weedkillers

Humus

Decayed organic material: animal manure, compost, leaves, etc. are all sources of humus which is a vital element in a fertile soil.

Hybrid

A new plant which is the offspring of two different species. This is usually indicated by an x between the botanical names of the parent plants. For example, *Viola* x *wittrockiana* (Pansy). (*See also* F_1; F_2.)

Inorganic

A chemical compound or fertilizer which does not contain carbon.

Insecticides see Sprays

Lime

Lime is useful for increasing the fertility of the soil and improves the texture of clay soils as it makes the tiny clay particles clump together. Hydrated lime should be applied at a rate of 3 oz to the square yard (75g to the square metre). If ground limestone is used it should be applied at the rate of 6 oz to the square yard (150g to the square metre). Lime should be applied during the late autumn and winter, i.e. between October and February. It must never be applied at the same time as other fertilizers or manure as it reacts chemically with them or prevents them from acting by making them less soluble. NOTE: Lime should never be given to ericaceous plants such as heathers, azaleas or rhododendrons.

Manures

There are two types: a) green manure and b) organic manure, both of which add humus to the soil.

Green Manure
This refers to any fast-growing crop which is planted specifically to be dug into the soil. Legumes, such as French beans or peas, are especially valuable for this purpose as they fix nitrogen which is vital for plant development. Italian rye grass can also be used as a green manure. Any plant used for this purpose must be dug in as soon as flowering starts.

Organic Manure
These are animal manures and include most of the nutrients which are essential to healthy plant growth. Farmyard and stable manures are perhaps the best known. Manure from poultry farms is not suitable for use in small town gardens as it tends to develop an offensive smell as it breaks down. Vegetable wastes are also organic manures and include spent hops and mushroom compost, leaf mould and, in coastal areas, seaweed. All of this material will have to be composted before use (*see* Compost). Also under the heading of organic manure come the processed manures such as bonemeal and dried blood which are very concentrated and are not used to treat the whole garden. As they are in powder form they add no humus to the soil. They should, however, be added to the soil when it is dug over by being sprinkled into the trench.

Monocarpic

Such plants survive for many years, but die after flowering.

Monocotyledon

Plants in this group have one seed leaf.

Mulch

A soft layer of material such as compost, other plant material or manure which is placed on top of the soil to conserve the moisture in the soil and to prevent the growth of weeds. Non-organic materials such as black polythene and stones can also be used as a mulch.

Native

An indigenous plant, i.e. a plant not known to have been introduced by Man.

Nutrients

Substances that provide the plant with nourishment including nitrogen, phosphorus, iron, potassium and magnesium. A soil deficient in these elements will produce poor plants. The three nutrients most essential to the health of the plant are nitrogen, potassium and phosphorus.

Organic

Refers to substances which are derived from the decay of living organisms and which therefore contain carbon.

Perennial

A plant which lives for more than two years and produces flowers and seeds annually throughout its lifetime. Examples are *Aster novi-belgii* (Michaelmas Daisy), *Campanula* spp (Bellflower) and *Gentiana acaulis* (Trumpet Gentian).

Mulching

Mulching with garden compost

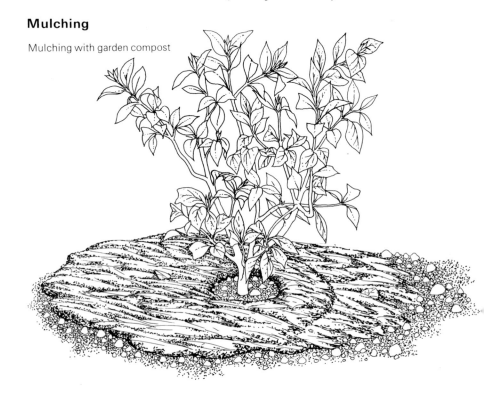

Mulching with black plastic

Mulching with stones

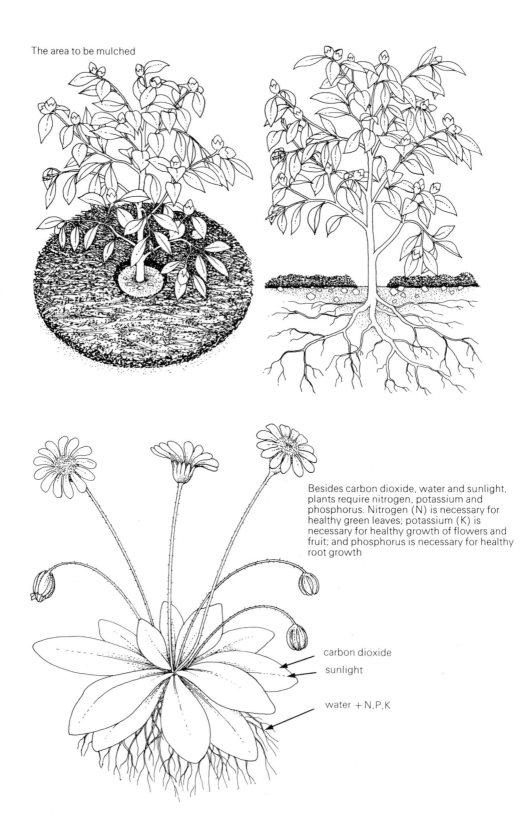

The area to be mulched

Besides carbon dioxide, water and sunlight, plants require nitrogen, potassium and phosphorus. Nitrogen (N) is necessary for healthy green leaves; potassium (K) is necessary for healthy growth of flowers and fruit; and phosphorus is necessary for healthy root growth

carbon dioxide

sunlight

water + N,P,K

Pots

Plants will grow in almost any container provided it has adequate drainage. However pots are generally made of clay, plastic or peat. Note that clay and plastic pots must be clean before use.

Clay
This is the traditional material for plant pots and clay pots are available in many sizes from a top width of 2 inches (5cm) upwards. Always soak a new clay pot in water for several days before using it. As these pots are porous they absorb moisture and consequently the soil in them dries out more quickly than in other types of pot.

Peat
These pots, made from Irish moss peat, are becoming increasingly popular. They are available in several sizes and are ideal for raising seedlings or cuttings as the pot can be planted out with its contents. The roots can grow through the walls of the pot when the plant is transferred to its final position and the pot eventually decomposes adding humus to the soil.

Plastic
These pots are very popular as they are light and easy to clean. They are also less likely to break than clay pots. As they retain moisture, there is a danger of overwatering.

Hanging baskets made of wire are very useful containers for trailing plants. Other suitable containers include tubs and troughs of various shapes and sizes.

Planting a hanging basket
Line basket with sphagnum moss
Half fill with compost
Push plants in spreading the roots out
Insert plants at the top of the basket and water in

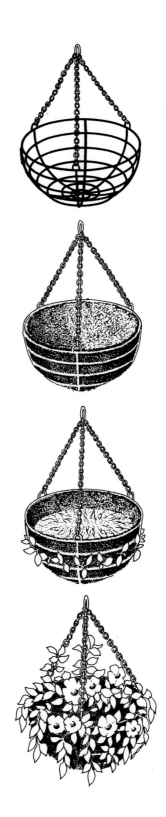

Types of pot

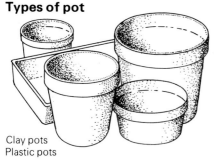

Clay pots
Plastic pots

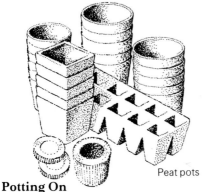

Peat pots

Potting On

The transference of a pot plant to a pot of a larger size than the one in which it is currently growing. It is time to pot on when the plant's roots begin to grow through the drainage hole.

This is done by watering the plant, removing it from the old pot together with the compost in which it is growing, and placing the plant and compost in the new pot which is then filled with compost to just above the original compost level. Firm down the compost and water the plant again. (*See also* Growing Medium; Pots.)

Pricking Out

The transference of seedlings from the pots or boxes in which they germinated to larger pots or boxes in which they have more space to grow on before being planted out in their final position.

Ray Floret see Floret

Rib

A leaf's main vein.

Root

That part of the plant, usually concealed underground which absorbs moisture and nutrients from the soil and which anchors the plant in the soil.

Seed

The reproductive unit of a flowering plant which contains the embryo plant.

Seed-bed

A seed-bed is made during the spring on soil that was dug and manured during the previous autumn. First fork over the soil. When it is dry firm it down and rake it lightly. Make a shallow drill using the edge of a hoe, ensure that the drill is straight by using a garden line (*see* Drill; Tools).

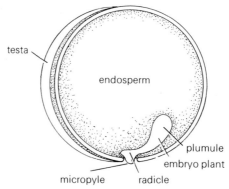

testa

endosperm

plumule

embryo plant

Cross-section of a seed

micropyle   radicle

The stages of potting on

Pricking out

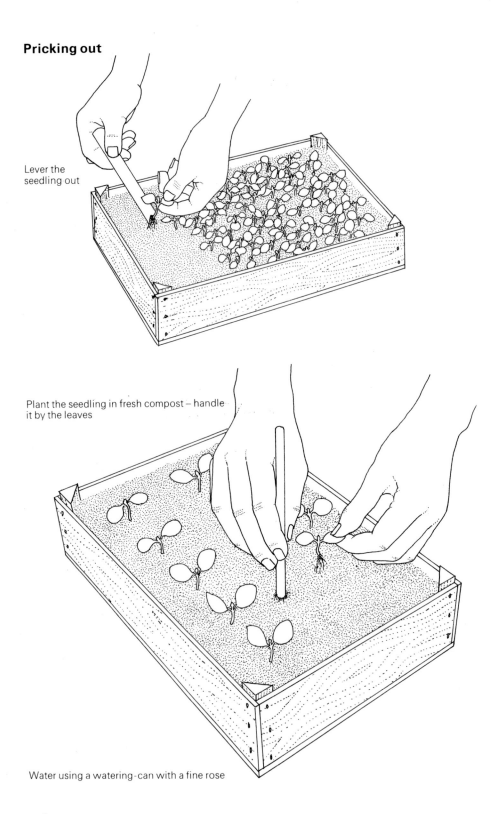

Lever the
seedling out

Plant the seedling in fresh compost – handle
it by the leaves

Water using a watering-can with a fine rose

Making a seed-bed

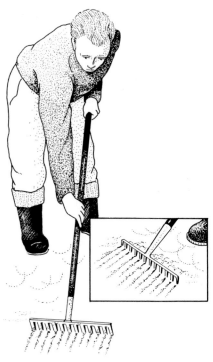

Firm the soil; rake to a fine tilth

Sow seeds evenly along the drill

Mark line to be drilled and scrape out a drill using the corner of a hoe or a rake

Cover the seeds by raking soil over them

Sow the seeds spreading them thinly along the drill. Then gently draw the soil back to cover the seeds using the hoe. Label the row before removing the garden line.

Seed Chipping

The process of filing or cutting the outer skin of a very hard seed, e.g., Sweet Pea, in order to allow moisture to enter, hastening germination.

Seedling

A young plant after germination which has a single unbranched stem. This term is also applied to an older plant which has been raised from seed.

Diagram of a seedling

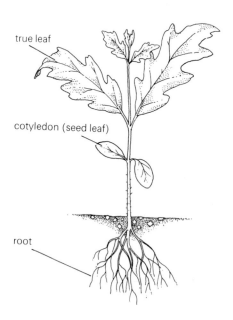

true leaf

cotyledon (seed leaf)

root

Shrub

A perennial plant which is smaller than a tree. It has branches and woody stems with little or no trunk.

Single Flowers

Flowers with no more than the normal number of petals.

Soil

You can make a rough estimate of the type of soil in your garden by carrying out the following test. Take a handful of soil and put it in a jam jar which is half full of water. Shake the soil and water together and leave the jar to stand for an hour or two. Sand and gravel will sink to the bottom, loam and clay will be suspended in the middle and humus will be floating on top. The proportions of the various materials will give you some idea of the type of soil you have. You can also tell a great deal about your soil by observing what happens when it rains. If the water stays on the surface for a long time it is likely that you have a clay soil. On the other hand, if the rain drains away quickly you are likely to have a sandy soil.

Soil tends to fall into one of the following groups, although you may find that your garden contains patches of different types of soil.

Chalk soil see Alkaline soil below

Clay soil
A dense heavy soil which tends to be wet, and is often water-logged during the winter. When you pick up a handful and roll it between your fingers it forms a solid ball. The tiny particles are tightly packed and plants can suffer badly during the summer as their roots are unable to penetrate the topsoil to obtain water from the subsoil. This type of soil can be improved by the addition of organic matter such as peat, leaves and garden compost. Lime can also be added (see Lime). In extreme cases land drains may have to be laid in order to drain the garden.

Soil

Soil separated out into its constituent parts

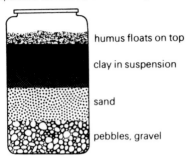

humus floats on top

clay in suspension

sand

pebbles, gravel

Sand trickles through the fingers.

Loam forms small crumbs

Clay forms a ball

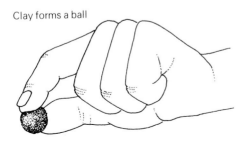

Loamy soil

A balanced blend of the other soil types containing a mixture of sand, clay and humus. When it is rolled in your hand it forms small crumbs. This is the ideal garden soil. The clay content prevents it drying out, the sand content ensures an open texture, and humus ensures a good supply of plant nutrients. Loams can be subdivided into sandy loams, which contain a high proportion of sand; heavy loams, which contain a high proportion of clay; and marls, which contain a high proportion of chalk.

Sandy soil

A type of soil found mainly in coastal areas. When you pick up a handful and run it between your fingers it trickles through them. It drains very rapidly and as a result nutrients are leached out of the soil. In order to prevent this happening bulky organic material such as farmyard manure, garden compost, or grass clippings should be added to the soil.

Soils can also be divided according to their acid or alkaline content. This is discovered by measuring their potential hydrogen (pH). Seedsmen can usually supply small kits for testing the pH of garden soil. These kits contain small bottles of chemicals which are mixed with small samples of soil following the instructions given with the kit and matched to a colour chart which gives the pH. Alkaline soils have a pH of 7 and above, neutral soils have a pH of 6.5–7; and acidic soils have a pH of less than 6.6. Soils are described as very acid when their pH is 4.5 or less. Most plants grow best in soils which are slightly acidic or are neutral, i.e. soils with a pH of 6.5–7.

Acidic soil

Such soils have a pH of less than 6.6. They can check the growth of some plants and very few plants will grow in a very acidic soil. Acidic soils can be corrected by the addition of lime, but the overuse of lime can be disastrous taking up to 15 years to correct. Peaty soils are acidic.

Alkaline soils
Such soils have a pH of 7 or over. They are chalky soils, i.e. they contain carbonate of lime in the form of chalk. It is difficult to correct this type of soil as rain washes more lime into the soil from the underlying chalk. Adding peat, leaf mould, grass cuttings, etc., will gradually make the soil more acidic, but the process takes a long time. Alkaline soils tend to suffer from a lack of nitrogen and potassium, both of which are essential to healthy plant growth.

Sowing

Fill a box or pot with compost, level it off and firm it down. Sow the seeds as thinly as possible: fine seed can be sprinkled from a piece of folded paper; sow large seeds individually in holes made with a pencil or a dibber.

Species

A sub-division of a genus, abbreviated to sp in the singular and spp in the plural. (*See also* Genus)

Sprays

Chemical sprays are of two kinds: insecticides and fungicides. All should be treated with great caution for as well as killing insects and fungus diseases they can injure human beings and domestic animals, sometimes with fatal consequences. Always use a sprayer with a nozzle which will produce a fine spray pattern with good coverage over the plants concerned. Always read the instructions on the

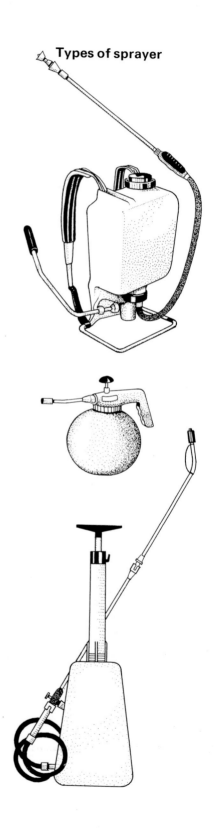

Types of sprayer

Sowing

Fill tray with compost; level off

Firm evenly; water

Make holes for large seeds using a dibber; sprinkle fine seeds on the surface using a folded piece of paper

Sprinkle compost on top; water using a watering-can with a fine rose

packet or bottle and *follow* them. Never mix chemicals together unless the manufacturer's instructions on the packet or bottle suggest doing so. There are an enormous number of chemical sprays available which are divided into two types according to the way in which they work. Systemic sprays are carried in the sap to all parts of the plant and will therefore affect pests or diseases in any part of the plant. A non-systemic spray is not carried around the plant and therefore kills pests by poisoning the surfaces on which they feed.

Rules for Spraying
1. Always read and follow the manufacturer's instructions.
2. Always wash off accidental splashes immediately and always wash your hands after using a spray.
3. Always keep containers well stoppered, ensure they are correctly labelled, and lock them up.
4. Always put empty containers in the refuse bin.
5. Always wash the spray equipment immediately after use, and never use spray equipment for any other purpose.
6. Never inhale the vapour.
7. Never harvest crops until the period of time given in the instructions has elapsed.
8. Never spray open flowers during the day. This will prevent helpful insects such as bees being killed.
9. Never spray when there is a breeze.
10. Never mix up more solution than you need – it cannot be kept.
11. Never spray the whole garden – spray only those plants that are being attacked by pests or disease.

Stopping

The removal of the growing tip in order to encourage the plant to branch out and become more bushy or in order to control the size or blooming of the flowers.

Subsoil

The soil below the fertile top layer of the soil.

Succulent

Any plant which has thick fleshy leaves and/or stems in which water can be stored. Such plants are adapted to survive in an arid environment.

Taproot

A thick fleshy root that descends for a considerable distance into the soil. Other roots branch off from it. Taproots can become very fleshy as with the carrot and parsnip.

Tendril

A slender clinging stemlike organ which is sensitive enough to twine around anything that it touches.

Tilth

The fine crumbly surface layer of the soil.

Topsoil

The fertile top layer of the soil.

Tools

Any job is more easily carried out using the correct tools. The tools that a gardener needs are a spade, fork, hoe, rake, trowel, dibber, garden line, secateurs, hose, watering-can and wheelbarrow. It is best in gardening,

Tools

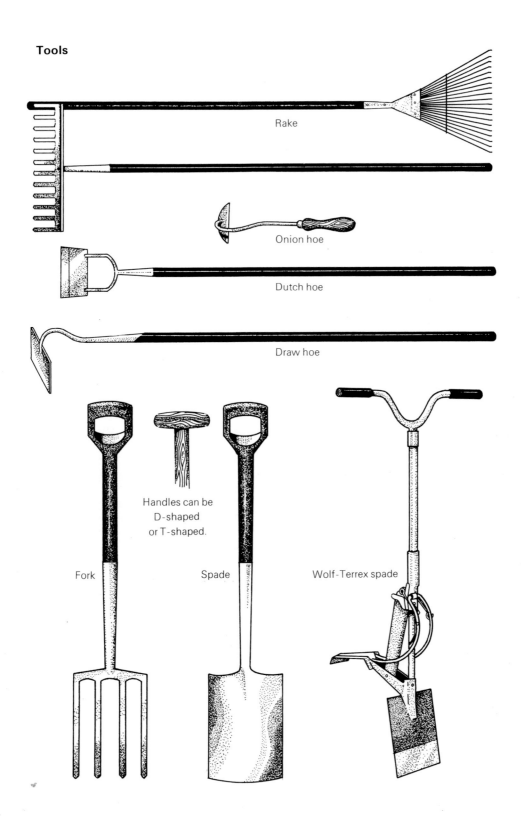

Rake

Onion hoe

Dutch hoe

Draw hoe

Fork

Handles can be
D-shaped
or T-shaped.

Spade

Wolf-Terrex spade

as in any other task, to buy the best quality tools you can afford. Tools with stainless steel blades are quite expensive but they are unlikely to rust and thus will give good service for longer than other types.

As far as caring for garden tools is concerned, they should always be cleaned after use and stored in a dry place where they should be hung up out of the way – to prevent damage to both the tools and yourself.

The handles of garden tools must be comfortable. Spades and forks are available with either a 'T' or 'D' shaped top to their handles. Most garden tools are available with wooden or polypropylene handles.

Dibber, Dibble
Any blunt pointed stick can be used as a dibber but it is more comfortable to use if it has a handle. It is used for making holes in the soil when transplanting seedlings.

Fork
Like the spade, (see below) the fork is available in two sizes (standard and border). It has four tines and is invaluable for breaking up lumpy soil, for lifting garden crops, and for shifting compost.

Garden line, Hoeing line
Two sticks with a length of cord tied between them are adequate for marking a straight line for hoeing or making a seed drill, but a ready-made line is easier to use. The line must be long enough to span the width of the plot. First one stick is pushed firmly into the earth, the line is unravelled until it reaches the other side of the plot, it is then pulled taut and the second stick is pushed firmly into the earth.

Hoe
The Dutch hoe is used to break up the surface of the soil to dispose of small weeds. It has a 'D' shaped or flat-bladed head and is used while the gardener walks backwards so that he does not tread on the ground he has just hoed.

The Draw hoe has a rectangular-shaped blade which is fixed at right angles to the handle. It is used to dispose of larger weeds and is used while walking forwards.

The Onion hoe has a short handle and must therefore be used while kneeling if the gardener is not to damage his back. It is very useful for working close to plants.

Hose and Watering-can
Both are essential. The hose is used for giving the plants a thorough soaking, the watering-can is used for selective watering and for applying liquid fertilizers and weedkillers. The watering-can should have a capacity of about 2 gallons (9 l) and both a coarse and fine rose are essential. If you intend to use a watering-can for spraying weedkiller, it is safer to have a can for holding spray materials only. A hose reel will prevent the hose becoming impossibly tangled during storage.

While on the subject of equipment related to water we must mention the water butt in which rain water can be stored for use during a drought.

Rake
This is used for levelling the soil, working in fertilizers and pulling out weeds. It usually has 12 teeth, but rakes with wide heads and more teeth are available. A fan-shaped wire rake is useful for removing cut grass and leaves but is not an essential tool unless you have a very large expanse of grass or are surrounded by trees.

Secateurs
A good quality pair of secateurs is essential. They are used for pruning and should be light and easy to use.

Tools

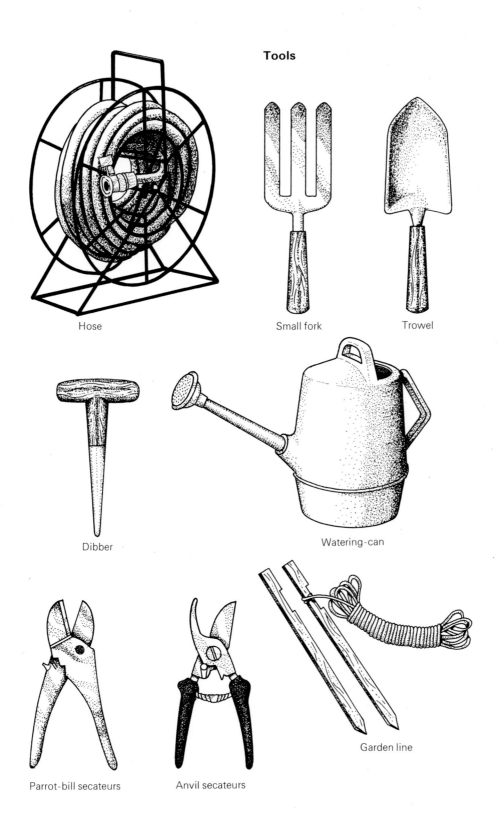

Hose

Small fork

Trowel

Dibber

Watering-can

Parrot-bill secateurs

Anvil secateurs

Garden line

Types of wheelbarrow

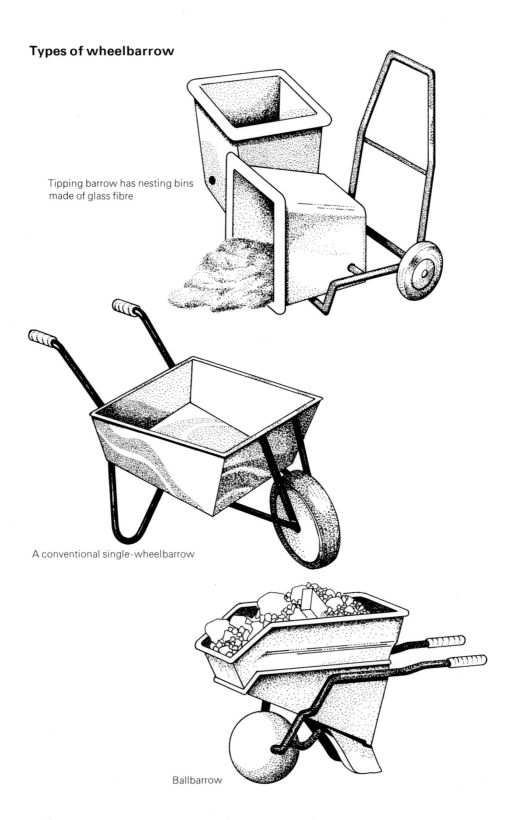

Tipping barrow has nesting bins
made of glass fibre

A conventional single-wheelbarrow

Ballbarrow

The blades must be sharp and hard-wearing. When using them ensure that the cutting blade is on top.

Two types of secateurs are available—the anvil type, which has a single cutting blade which cuts against a broad 'anvil' type blade, and the parrot-bill type, which has a scissors action.

Spade
Two types are available, the standard size, which has a blade measuring 7 × 11 inches (18 × 28cm), and the border size, which has a blade measuring 6 × 9 inches (15 × 23cm). The smaller size is easier to handle and is usually perfectly adequate for the small garden. A modern development which has made digging easier for people with weak backs is the semi-automatic spade, e.g. the Wolf-Terrex.

Trowel
This tool is used for planting out and weeding. If you can remember its overall length and the lengths of the blade and the handle it can be used as a rough guide to spacing plants when planting out. Forks are available in the same sizes as trowels and these tools are also available in miniature form for use with houseplants.

Wheelbarrow
The barrow must be sturdy as it will be used for moving all types of material about the garden.

Transplanting

The process of moving young plants from one place to another to give them more space in which to develop. When plants are moved it is essential to ensure that they are firmly planted in their new position. Test this by pulling gently on a leaf; if the whole plant moves it is not planted firmly enough.

Tree

A woody plant with a central main trunk from which branches radiate.

Variegated

A leaf which is marked by regular or irregular stripes in a colour different from that of the leaf or by patches of another colour, such as white, cream or yellow. Such plants are usually grown for their foliage as their flowers tend to be insignificant.

Variety

A group of plants which vary from the species type. It may also refer to a cultivar or a member of a hybrid group. (*See also* Cultivar; Hybrid.)

Vegetative

This refers to methods of propagation other than by seeds, for example by cuttings, layering, root division or grafting.

Weedkillers

Weedkillers are available in three types: selective weedkillers, mostly used for weeds in lawns; contact weedkillers, which kill all parts of the plant with which they come into contact; and residual weedkillers which remain near the surface to kill weed seedlings as they emerge. Weedkillers should not be used near fruit and vegetable plants nor should they be used where plants are growing close together. Remove weeds from such areas by hand. Unlike insecticides and fungicides, herbicides can be applied using a watering-can. Like all types of chemical spray they should be used on a day when there is no breeze. Ensure that the watering-can is washed well after use.

ACKNOWLEDGEMENTS

The publishers would like to thank Jonathon Bosley for general photographic help, and the Harry Smith Horticultural Photographic Collection for permission to reproduce the following illustrations: half-title page, frontispiece, 40, 62, 86, 104, 145 (right), 148, 160, 172 and 173.

All other illustrations Copyright © Suttons Seeds Limited, whose help has been invaluable in the compilation of this book.

Acknowledgement is also made to Sheila Ladner for her invaluable assistance in the production of this book.